MODERN RADIO SCIENCE 1996

International Union of Radio Science

Modern Radio Science 1987
Edited by A. L. Cullen

Modern Radio Science 1990
Edited by J. Bach Andersen

Modern Radio Science 1993
Edited by Hiroshi Matsumoto

Modern Radio Science 1996
Edited by J. Hamelin

Review of Radio Science 1984–1986
Edited by G. Hyde (Available from URSI Secretariat)

Review of Radio Science 1987–1989
Edited by G. Hyde (Available from URSI Secretariat)

Review of Radio Science 1990–1992
Edited by W. Ross Stone

Review of Radio Science 1993–1995
Edited by W. Ross Stone

MODERN RADIO SCIENCE 1996

Edited by

J. HAMELIN

Head of Space Division
Ministry of Information Technology and Post Affairs
Paris, France

Published for the
International Union of Radio Science
by
Oxford University Press
1996

This book has been printed digitally in order to ensure its continuing availability

OXFORD
UNIVERSITY PRESS

Great Clarendon Street, Oxford OX2 6DP

Oxford University Press is a department of the University of Oxford.
It furthers the University's objective of excellence in research, scholarship,
and education by publishing worldwide in

Oxford New York

Auckland Bangkok Buenos Aires Cape Town Chennai
Dar es Salaam Delhi Hong Kong Istanbul Karachi Kolkata
Kuala Lumpur Madrid Melbourne Mexico City Mumbai Nairobi
São Paulo Shanghai Singapore Taipei Tokyo Toronto

with an associated company in Berlin

Oxford is a registered trade mark of Oxford University Press
in the UK and in certain other countries

Published in the United States
by Oxford University Press Inc., New York

Reprinted 2001

A catalogue record for this book is available from the British Library

Library of Congress Cataloging in Publication Data
(Data available upon request)

URSI edition ISBN 0-19-856529-1
OUP edition ISBN 0-19-856530-5

Preface

Published every three years when the General Assembly of the International Radio Science Union (URSI) takes place, this book presents a wide range of the most recent and outstanding activities in the fields of radio, telecommunication and electronic sciences in particular "the scientific aspect of telecommunications using electromagnetic waves, guided and unguided, as well as the generation and detection of these waves and the processing of signals embedded in them".

The present book includes three lectures considered to be salient features of the last years work in electromagnetism and tutorials suggested by the commissions of the Union. Those tutorials stand out to be characteristic of the recent commission activities.

Many of the topics in this book partake of up to date economical activities : mobile communication applications, numerial methods associated, electromagnetic protection of very large systems and health point of view of personal communication services. It testifies of the growing role of radioscience in the emerging information society.

Other contributors bring us towards the confines of the universe, physics or measure. Every topic points out of the extraordinary dynamism of this scientific domain, of its constant renewal, of its astonishing extension of applications.

This book like the previous ones, is a synthetic view that will allow the scientists, whether they are specialists or not, to apprehend the lastest steps in Radio Science research.

Finally, I would like to thank all the contributors for their considerable efforts in preparing their manuscripts.

J. HAMELIN

Paris, May 1996

Contents

General Lectures

Tutorial Lectures

Contributions

General Lectures

Dr R. K. Raney
Applied Physics Laboratory, The Johns Hopkins University
Laurel Maryland 20723, USA

Prof. W. Lauterborn and Dr U. Parlitz
Drittes Physikalisches Institut, Universität Göttingen
37073 Göttingen, Germany

Dr M. Joindot
France Télécom/CNET
22307 Lannion, France

Tutorial Lectures

Commission A
Prof. Dr. J. Niemeyer
Physikalisch-Technische Bundesanstalt
Bundesallee 100, 38023 Braunschweig, Germany

Commission B
Prof. R. Tiberio[1], Prof. S. Maci [2] and Prof. A. Toccafondi [1]
[1] *University of Sienna*
via Roma 56, 53100 Sienna, Italy
[2] *University of Florence*
via S. Marta 3, 50139 Florence, Italy

Commission C
Prof. R. L. Pickholtz
The George Washington University
Phillips Hall 6 Floor, Washington, D.C., 20052, USA

Commission E
Dr J.P. Parmantier [1] and Prof. P. Degauque [2]
[1] ONERA
8, rue des Vertugadins, 92190 Meudon, France
[2] Université des Sciences et Techniques de Lille 1
59655 Villeneuve d'Ascq, France

Commission F
Dr K. H. Craig
Rutheford Appleton Laboratory
Chilton, Didcot, Oxfordshire OX11 0QX, UK

Commission G
Dr B. M. Reddy
National Geophysical Research Institute
Uppal Road, Hyderabad, 500007, India

Commission H
Prof. D. B. Melrose
School of Physics, University of Sydney
Sydney, NSW, 2006, Australia

Commission J
Dr J. M. Moran
Harvard-Smithsonian Center for Astrophysics
60 Garden Street, Cambridge, MA 02138, USA

Commission K
Prof. M. A. Stuchly
Department of Electrical and Computer Engineering, University of Victoria
MS 8610, Victoria, B.C., V8W 3P6, Canada

FROM COHERENCE TO CONFUSION: A CONSERVATIVE SAR VIEW

R. Keith Raney

1 Introduction

Resolution is a measure of image sharpness, whereby two closely spaced objects yield echoes that are separable. Imaging radars are distinguished one from another primarily by the means used to achieve resolution in the azimuth (cross-range) direction, for which the available data consist of receptions from many transmitted pulses. A real aperture radar (RAR), or side-looking airborne radar (SLAR) in some references, has access only to the power of the received signals. Processing strategies cannot materially enhance their azimuth resolution. Therefore a real aperture radar's azimuth resolution is bounded below by the azimuth width of the antenna pattern, which is proportional to range. On the other hand, a synthetic aperture radar (SAR) preserves the inter-pulse phase structure of the received signals. The SAR (image) processor uses this phase information to enhance azimuth resolution, a process known as focusing. The resultant resolution is much smaller than the antenna pattern width. Indeed, SAR azimuth resolution is independent of range, to first order. Stated more formally, the SAR signal ensemble has large azimuth time-bandwidth product (TBP_A). The theoretical resolution is equal to the inverse bandwidth (in appropriate temporal or spatial units), which, for conventional narrow-beam side-looking geometries, is approximately equal to half of the antenna aperture size in the azimuth plane. For satellite SARs, TBP_A is on the order of 1000. Usually, linear frequency modulation or similar encoding is applied to each of the transmitted pulses, so that the range dimension also has large time-bandwidth product, TBP_R.

Enhanced resolution is achieved at a cost. A synthetic aperture radar system must have: 1) sufficient stability (coherence) to retain the correct signal phases; 2) a memory to store the appropriate sequence of echoes; and 3) a processor to do the work of image formation. The processor must integrate over TBP_A x TBP_R input elements to determine each output image value. Fortunately, frequency domain algorithms can introduce considerable efficiency into these operations.

In general, for a given resolution, image quality is improved with increased averaging. Averaging occurs naturally for real aperture radar data, but must be implemented explicitly in the processor for a SAR. Radiometric image quality is proportional to the number of statistically independent looks (degrees of freedom) divided by the product of the range and azimuth resolutions.

2 Fundamental SAR Principles

Popular opinion has it that SAR is a difficult specialty, perhaps designed with devious intent to baffle newcomers. One could easily conclude that there are too many aspects to consider, including a few tricks known only to those versed in the art. On the contrary, SARs are well behaved. Synthetic aperture radars are elegant systems whose fundamentals may be summarized rather compactly. Six basic principles are sufficient to derive most important aspects of SAR. Stripped of its more familiar appearance, the essential features of a SAR system may be compactly modelled as one "black box", which in effect is a microwave transducer that converts an extensive complex input field into a suitable output mapping. This transducer implies two sequential operations: data collection, the task of the radar; and image formation, the task of the processor. The observation space (scene) is probed by a microwave signal, to produce the reflections that are collected by the radar. The output space (image) is a distribution of reflected power. Imagery usually has two dimensions, range and azimuth. The *Magellan* imagery of Venus is a most dramatic example from this class of radar [*Johnson,* 1991; *Pettengill,* 1993].

To provide a workable basis for the rather formal treatment of this section, it is convenient to make certain realistic assumptions. These are: 1) the system uses only one transmit and receive polarity; 2) input signals from the scene combine linearly; 3) signal manipulations on the output are linear in power; and 4) the system uses a magnitude squared method of detection. The single polarimetric channel admits a scalar model, which simplifies both the mathematics and the attendant discussion. The scalar results extend readily to full polarimetric systems. The assumption of input linearity for a SAR is true outside of the radar, and almost always true inside [*Harger,* 1970]. Signal combination prior to detection is coherent. The dominant and essential nonlinearity is signal detection. Detection is the process through which certain radiometric information carried by the electromagnetic (em) field, such as the power of the radar echo, may be extracted. For SARs, this step takes place in the processor after the complex image is focused. Magnitude squared detection, one of several possible methods, is convenient for mathematical reasons, and finds justification in nature as well. Following detection, additional averaging may be performed, known as incoherent integration since it occurs in the power domain.

The principles introduced in this article establish that radar systems have an underlying logical order, which is appealing in its own right to applications specialists as well as to systems theoreticians. At a more practical level, these principles provide a relatively simple basis for understanding SAR system performance, which may be helpful when more detailed questions are encountered. The discussion that follows each of the principles is meant not only to provide a first order explanation, but also to serve as an enticement for the reader to probe more deeply into these and related concepts.

Principle 1. Coherent scene illumination. *Input to a remote sensing radar from the scene is a linear summation of voltages from reflecting elements that are illuminated by a coherent em wave field.*

This principle, while obvious at one level, immediately sets radar remote sensing apart from optical remote sensing. Most optical remote sensing systems use natural radiation that is polychromatic, and emitted from an extended source. The radiation has random phase structure between any two arbitrary points in the illuminating field. Such em fields are incoherent. This means that the underlying phases are random, which implies that the combinations of reflections from more than one scene element add in power.

In distinct contrast, for most radars the illumination from each transmitted pulse is essentially monochromatic, emitted from a point source. This implies that the radiating field has structured phase fronts, which may be well represented for most purposes by spherical surfaces, centered at the radar. Thus, the em illumination incident on the scene is coherent. This means that the reflections from more than one scene element combine through vector addition.

Imagery is generated by processing over a signal ensemble comprised of reflections from many scatterers and from many pulses. If the radar is controlled so that the pulse sequence is mutually coherent, then the phase structure within the reflections from each scatterer can be exploited during processing. This is the case with a SAR.

If any element of the scene has a reflectivity that changes during the radar observation interval, then the relative phase of the reflected field will change. A systematic phase change in the scene may be observable by a coherent radar, which is the basis for moving target detection and radar interferometry, for example. Random phase changes in the scene, in the radar, or in the propagation path may lead to loss of image resolution.

Principle 2. Image power mapping. *Image domain output from a remote sensing radar is a mapping of power, derived from scene reflectivity, and presented as a two-dimensional array of numbers.*

It is customary to think of a radar image as a visually observable entity, such as a photographic print, whose brightness (whiteness in a positive rendition) at each location is proportional to the strength of the corresponding radar echo. As was the case with the first principle, the statement seems obvious, but there are significant implications. No matter what kind of radar is being used, the microwave transducer must estimate the power from the sum of echo voltages for each location in the image. For coherent fields, the power is given by the magnitude squared of the vector voltage sum, not by the sum of the powers of individual echoes. Constructive and destructive interference takes place between the members of the signal ensemble. This leads to statistical variation of the image power estimates.

In the simplest scenario, the scene is composed of a set of stationary scatterers, and no phase errors are introduced through wave propagation between the scene and the radar. In this case, coherence within a group of signals is determined primarily by the coherence of the radar. SAR systems maintain phase coherence between multiple pulses, whereas for real aperture radars the pulses each have random reference phase, and so they are incoherent pulse to pulse. SAR processors usually are designed to introduce a controlled amount of incoherence, in order to improve image quality. The image power mapping from a SAR is described quite well by a partially coherent system model. The statistical properties of a SAR image are functions both of the scene and of the system.

Most SAR systems use digital processing, and hence their images may be presented as a discrete numerical array, in which the digital number at each datum point (pixel) is representative of image brightness. The image domain digital numbers may be converted to estimates of the scene's reflectivity coefficients through the applicable calibration rule. Note that pixel separation is bounded above by the Nyquist sampling limit. In a real image, at least two pixels are required to support the stipulated resolution distance, although often this constraint is relaxed because the image spectra are weighted.

Principle 3. Well behaved transformation. *The transformation from scene data to image data by a remote sensing radar may be characterized if, and only if, both the response to an isolated point scatterer, and the response to a uniform Gaussian distributed scattering field, are determined.*

This important principle applies to all partially coherent systems that may be represented with the mathematics of quadratic filters, a wide class that includes optical and radar sensors [*Raney*, 1968]. The definitive feature of such systems is that they operate on an input signal field in voltage to produce an output data field in power, using linear filters and square law detection, combined according to the degree of coherence in use by the system, and by the prevailing coherence of the scene.

It is well known that a linear filter may be characterized by its response to an arbitrarily short test signal, leading to the so-called impulse response function [*Woodward,* 1957]. Application of the Fourier transform to the impulse response function produces a frequency domain description of the linear filter, known as its transfer function.

It is also true that the frequency distribution of a linear filter may be determined by observation of its frequency response to white Gaussian noise [*Papoulis,* 1968]. The filter's impulse response may be found by inverse Fourier transformation of the frequency function. Thus, for linear systems, the impulse response and the system frequency transfer function are logical equivalents of each other. This simple relationship is not sufficient for radar systems.

For a SAR, as with all quadratic filters, measurement (or specification) is needed of both the impulse response and the frequency function [*Raney,* 1985]. The purpose of the impulse response test is to estimate the system imaging function, analogous to the point spread function encountered in optics. A SAR normally is operated in a partially coherent mode, as noted under principle 2. The purpose of the Gaussian response test is to estimate the extent of system partial coherence. It turns out that the Gaussian test also determines the system frequency distribution, which is useful in certain image analysis situations.

The essential lesson in this principle is that a SAR is well behaved in the best sense of system theory. Coherent input field summation in combination with the partially coherent operations of the radar and processor results in imagery that has features that are robust and predictable, even if their properties are less familiar than those of incoherent optical devices that behave linearly in power.

The following principles highlight two radiometric rules for SAR images that follow from the well behaved principle, and from the benefits of large time-bandwidth product.

Principle 4. Conservation of energy. *For any neighborhood, if the scene is a Gaussian field, and all available data are used in each case, then the tone of the corresponding area in the image is a constant, independent of system focus, system coherence, or scene coherence.*

A Gaussian field (mathematically speaking) is an important idealized scene for imaging studies [*Goodman,* 1976]. A wheat field (from the point of view of a farmer) is the classic practical example, having spatial features that are far too small to be resolved by the radar, and having a uniform average reflectivity over an area much larger than the resolution cell [*Brown*, 1967]. The radar is to estimate the power reflected from the wheat field by observing the corresponding average brightness value, or tone, in the image. The Gaussian property in this case derives from the fact that for each element in the image, there are many individual scatterers, such as the heads of the wheat shafts, contributing to the total scattered signal.

One might think that the tone would depend on the details of the field, including the number, size, and location of each scattering element. Mean reflectivity could be changed by random motion in the scattering field, due perhaps to a passing breeze. Further, knowledge of all radar and processor phase errors could be a reasonable requirement. Even more tempting, one might think that an error in system focus, or a change in the degree of partial coherence used in the processor, would lead to a change in the image tone. Wrong on all counts! Within very broad constraints, image tone is robust even when certain parameters change, which is the point of the compact wording of this principle.

The principle of conservation of energy is very powerful. Among other things, it establishes a solid basis for radiometric calibration of a SAR whereby image brightness values can be transformed to scene reflectivity estimates. For a given SAR, conservation of energy provides assurance that image average brightness (tone) is preserved independent of the number of degrees of freedom (looks) in the image data. Furthermore, given the same viewing geometry, polarization, wavelength and system bandwidth, two different (calibrated) radars should provide equivalent mean reflectivity estimates for the same scene. This fact allows measurements from scatterometers and other radars that are not "SAR systems" to provide reflectivity data that are directly applicable to SAR observations.

There is a higher order effect. One may show that the image energy actually is weakly dependent on SAR processor focus. For a focus error of δ, the image energy is proportional to $1/(1 + \delta/\mathrm{TBP})$. To put this into perspective, impulse response broadening begins to be observable for $\delta \sim 1$. Since SAR systems typically have large TBP, it follows that energy conservation is robust even in the face of small focus errors.

Principle 5. Conservation of confusion. *For any neighborhood, if the scene is a Gaussian field, and all available data are used in each case, then the uncertainty associated with each corresponding area in the image is a constant, independent of system focus, system coherence, or scene coherence.*

At each neighborhood in the SAR image of a uniformly reflecting random field, there is a statistical distribution of the brightness, even though the input scene has constant average reflectivity. This component of image texture is known as speckle [*Goodman*, 1976], which is one consequence of the coherent illumination of the radar, as noted in principle 1. To first order, there are three measures of speckle: its statistical variation at a given pixel; its correlation with adjacent pixels in range; and its correlation with adjacent pixels in azimuth. The expected change in radiometric response at each pixel usually is measured by its covariance. (The covariance often is called the variance in SAR literature, a source of potential confusion.) The covariance is defined as the square of the second moment with respect to the mean value. The normalized covariance, which is the covariance divided by the local mean squared, is a measure of the amount of averaging in the data. The statistical number of degrees of freedom (or "looks" in SAR parlance) is given by the mean, squared, divided by the standard deviation.

Speckle radiometric variation at any image pixel is correlated with that at neighboring positions. The width of the covariance function is defined as speckle correlation length. The speckle correlation length is equal to the inverse of the coherent bandwidth of the system, which is true in both the range and the azimuth dimensions.

The uncertainty at each point is measured by the volume under the normalized two-dimensional image covariance function. The principle of conservation of confusion states that the uncertainty volume is a constant of the system, under a wide variety of circumstances.

The importance of this may be illustrated by two special cases. First, the width of the covariance function (or equivalently the width of the peak of the auto-correlation function) is equal to the inverse system bandwidth in each of the two image dimensions. As is true for any imaging system, the width of the impulse response is reduced as the focus is adjusted towards the ideal setting. At perfect focus, the impulse response width is minimized. This minimal width is defined as the system (ideal) resolution, which is equal to the width of the speckle covariance function. For a SAR, the covariance function is essentially invariant under a change in system focus. This means that the covariance function may be used to estimate the system bandwidth. This establishes the potential system resolution, which serves as an independent check on the results of an impulse response test. Also, this approach provides a practical means to estimate the system response function using real data from an arbitrary scene.

Second, a corollary of this principle states that there is a trade-off between resolution and the amount of averaging that may be used to reduce speckle variance [*Porcello et al*, 1976; *Moore*, 1979]. In brief, the ratio of the number of (statistically independent) looks divided by the product of the corresponding resolutions is a constant of the system, which is proportional to its (two-dimensional) Shannon channel capacity. System bandwidth may be used either to maintain fine resolution (the fully coherent case), or to generate statistically independent images to be averaged together to form the final image (the partially coherent case). For the latter, the reduction in image variance is equal to the proportionate increase in resolution cell size. A decrease in radiometric uncertainty implies an offsetting increase in spatial uncertainty. Uncertainty, or confusion, is conserved. This consequence is an example of the uncertainty principle for energy conservative systems [*Papoulis*, 1968].

This principle states that multi-look processing is information conservative when confronting diffuse scattering from area-extensive objects. This should not be surprising : information is bandwidth, and the statement of the problem does not admit that the two-dimensional system bandwidth should change. The trade-off is between radiometric resolution and spatial resolution. For distributed diffuse scattering, the trade-off is a toss-up. For point targets, however, for which resolution has merit in its own right, and there is less statistical variation to suppress, the trade-off is always in favor of resolution (and fewer looks), up to the limit dictated by either the linearity of the system or the coherence of the scene and the propagation path [*Brown and Palermo*, 1965; *Raney*, 1968].

There is a higher order dependence on processor focus, similar to that found above. One may show that the normalized covariance is proportional to $1/(1 + \delta/\text{TBP})$. Since SAR systems typically have large TBP, it follows that the image covariance is robust in the face of small focus errors.

The principle of conservation of confusion is a natural counterpart to the principle of conservation of energy. Together, these two principles describe the fundamental SAR performance of the first order image statistic, tone, and the second order image statistic, texture, in response to a uniformly reflecting distributed scene. These radiometric principles are complemented by one that undergirds a SAR's spatial performance.

Principle 6. Conservation of coordinates. *For a range/Doppler radar, the output mapping has constant range and azimuth coordinate locations on the imaged surface, independent of angular rotations of the sensor.*

This principle follows from the first and third principles, and the illumination geometry customarily used with SAR systems. It implies that there is a fundamental spatial mapping accuracy available from a range/Doppler radar such as a SAR that is inherently less sensitive to sensor pointing precision than for most other imaging systems. For example, optical instruments derive mapping information from offsets relative to the angular orientation of the sensor itself. Mapping accuracy for those sensors is limited by the angular pointing stability of the sensor, and degrades in proportion to the distance of the scene from the instrument.

For range/Doppler radars, the situation is very different. In range, the coordinate is proportional to time delay, which is not affected by sensor rotations. In azimuth, the coordinates are established by the Doppler frequency gradient across the scene, which in turn is determined by the relative vehicle-target velocity, and its the radar's instantaneous along-track position. In either dimension, an error in sensor pointing does not impact spatial location of the coordinate system with respect to the sensor.

3 SAR Processing

The history of SAR rests primarily on the history of SAR processing. From the processor's point of view, a good synthetic aperture radar is expected to do very little other than to collect the signal data coherently, and to deliver the data faithfully to memory, whose accumulated contents serve as the input to the processor. It follows from first principles that the data to be processed is a linear combination of signals from individual scattering elements. Hence, to understand SAR processing, it is necessary and sufficient to consider the case for only one representative signal, namely, the output of the radar in response to a single point scatterer.

Azimuth SAR processing may be viewed as an exercise in wave front analysis. The problem is non-trivial because 1) each received signal depends on the parameter of range which may vary substantially across the intended field of view, and 2) the signal is coupled between the range and azimuth directions. SAR processing algorithms now in use may be studied through the means used to cope with these two complications [*Bamler*, 1992]. Early SAR processing ignored wave front curvature effects, and reached a temporary performance plateau with coherent optical devices. Whereas the optical analogy once provided insight into focusing SAR signals, now the optical construct is a limiting framework. Whether or not there is conscious awareness of it, most investigators are in the grips of a mind set that limits as well as guides creative energies. Such a mind set has been described as a paradigm [*Kuhn*, 1970]. Posed as a generalized two-dimensional signal transformation problem, SAR processing possibilities exceed those to be seen from an optical perspective, no matter how implemented.

3.1 SAR processor history: an interpretation

The story of synthetic aperture radar (SAR) begins with the introduction of the concept by Carl Wiley in June, 1951 [*Wiley*, 1985]. He observed that for a moving coherent radar, reflections from objects are characterized by a Doppler shift unique to their angular orientation with respect to the velocity vector of the vehicle carrying the radar. Wiley reasoned that spectral analysis of the recorded signal histories should reveal each scatterer's azimuth location. Spectral precision is inversely proportional to record length, which itself is proportional to the scatterer distance (range) from the radar velocity vector. Azimuth spatial resolution is the product of angular resolution and range. Wiley concluded that the two range dependencies should cancel out: spectral analysis of a coherent signal history implied that azimuth spatial resolution should be constant as a function of range. His observations have proved to be correct.

The early days

The first demonstration of SAR processing was at the University of Illinois in 1952 [*Sherwin et al*, 1962]. Spectral analysis of the radar signals was implemented using the electronic tools available at the time. These included recirculating delay lines and storage tube techniques. The results were sufficient to establish the feasibility of Wiley's idea, but were impeded by the limitations then imposed by technology. In parallel, experiments using coherent optics were taking place at the Willow Run Laboratory of the University of Michigan [*Cutrona and Hall*, 1962; *Cutrona et al*, 1960]. Their approach was to apply known optical configurations to implement explicit analog correlation of a reference function against the ensemble of radar signals that had been rendered onto a film transparency. The original correlator relied on a reference function that emulated the Doppler modulation expected from a point scatterer, namely, a linear frequency modulated (fm) signal whose length and rate of frequency change were direct functions of radar range. The approach led to the first demonstration radar, the UPD-1. Using an off-line electronic analog computer, the reference signal was painstakingly constructed on photographic film as an opaque and clear mask. The correlator used at this stage may be viewed as a logical (and Fourier transform dual) alternative to spectral analysis, and constituted an approach which, at the time, was capable of better performance than that of electronic devices. Furthermore, the optical correlation stage was very fast, although the end-to-end process was encumbered by the need for film development.

Meanwhile, also at the Willow Run Laboratories, a young master's degree student was working on alternative representations of the SAR processing problem. Emmett Leith made the important observation [*Leith and Upatnieks*, 1965] that the reference function was equivalent to a one-dimensional Fresnel zone plate, which had been known in optics since the early nineteenth century [*Born and Wolf*, 1959]. Furthermore, Leith noted that the zone plate was equivalent to the phase structure of a spherically radiating coherent field as it impinged on a planar surface. Leith's next observation was the key: if the required reference function were equivalent to the phase structure of a spherical wavefront, then the corresponding scatterer wavefront should be able to be focused without recourse to a reference function [*Leith,* 1971]. He proved his point directly. In effect, he went into the correlator laboratory, removed the reference function, rearranged the optics, and a better image resulted! The age of optical processing for SAR data was born.

A paradigm established

Perhaps of more significance, the compelling analogy between SAR and conventional optical imaging systems was recognized for the first time [*Cutrona et al,* 1966]. The optical analogy is a strong one, and continues to pervade the field of SAR processing. Leith's contribution established a working paradigm in the field. Attention turned from spectral analysis, and focused on optical implementations. As a result, tremendous strides were made. It was noted that a linear fm range pulse had the same kind of focal properties as the azimuth Doppler modulation. Thus, optical processors were developed that performed simultaneous beam compression and pulse compression [*Leith,* 1971]. Of course, the respective fm rates were different, which in optical terms meant that the range and azimuth focal lengths were not comparable. Furthermore, the fm rate of the azimuth dimension varied with range. To match this, the azimuth focal length of the optics would have to be a function of range, so special optics were needed [*Cutrona et al,* 1966]. Conical as well as cylindrical lenses were required in early processors of this type. Eventually, in the mid-1960's, the approach was perfected in the anamorphic telescopic tilted plane tracking processor developed by Leith, Massey, VanderLugt, and others [*Kozma et al,* 1972]. Diffraction limited imaging performance was achieved. The Michigan processor still represents the state of the art for film-to-film optical methods.

It should be observed that the approach initiated by Leith was radically different from those previously known. In particular, the UPD-1 optical processor of the late 1950's had been a true correlator, using an optical analog computer to explicitly perform a cross-correlation of the signal structure with a reference function, which had been made as a photographic transparency. The Leith approach [*Leith,* 1971] was based on an entirely different principle, holography, later to earn him honorable mention for a Nobel prize. The elegance of Leith's observation lay in his recognition that the image already existed; it remained only to bring its range and azimuth focal planes into coincidence, and at the desired location.

Warning signs

Most people working in SAR processing today recognize that the Fresnel zone plate simile has severe limitations. It models only the central part of the phase history that is characterized by linear fm. As the field of view expands, the correct modulation should represent a hyperbolic phase history. Additional complications come from the fact that a radar is both a ranging device as well as a coherent recorder of the azimuthal signal phase. For angles away from the zero-Doppler plane, the signal received from a given scattering element travels a greater distance than its reference range. This effect, known as range migration or range curvature, becomes important when the range shift exceeds the range sample width. Other subtleties arise for satellite SAR data. With a circular orbit over a rotating Earth, the phase fronts include range-dependent departures from the hyperbolic model, and the zero-Doppler plane is normal to the ground track of the satellite rather than to the spacecraft velocity vector. Precision SAR processing demands that all such range curvature characteristics must be fully compensated.

From an optical processing point of view, range curvature effects are a nuisance. Optical processors have been modified to account for modest range curvature, but only approximately [*Brown and Fredricks*, 1969; *Leith*, 1973]. Non-spherical optics, or holographic correction plates, are required to account for hyperbolic phase modulation terms.

In the late 1970's, development of digital processing techniques for radar remote sensing was supported in both the United States and in Canada for application to Seasat SAR data. The first image was published in Aviation Week (November 1978) shortly after conclusion of Seasat operation. The processing algorithm then used was closely analogous to operations inside of an optical processor. The principal difference was that all steps were taken in series, rather than in parallel. In the Canadian algorithm [*Bennett et al*, 1980], the signal was Fourier transformed in range, match filtered, then inverse transformed, leading to the range image and azimuth signal domain. Subsequently, the signal was azimuth transformed (range/Doppler domain), match filtered in azimuth, and inverse transformed. A problem arose, however. For Seasat parameters, range curvature is significant, exceeding 40 resolved range bins. The processor architecture had to be augmented to include range curvature correction, which was applied in the range/Doppler domain prior to azimuth focusing. This, and other aspects of the implementation, required digital interpolation of the complex data set [*Bennett and McConnell*, 1980; *Wu*, 1980].

Several approaches were proposed to compensate for range curvatures that exceeded the capabilities of the initial digital implementation. So-called secondary range curvature correction algorithms were designed [*Jin and Wu*, 1984]. Again, the phenomenon was treated as a serious imaging impediment, and required extraordinary processing effort to overcome it.

In short, range curvature was regarded as a nuisance in digital SAR processors, just as it had been in optical processing. Range curvature attracted attention only after the task of azimuth focusing had been addressed. One could argue that this was a consequence of viewing the SAR/processor as an optical analog. The paradigm of optical processing which constructively influenced early digital SAR processing had become a burden.

It should be acknowledged that the first digital SAR processors were substantial contributions in their time, and were essential, for example, to the evident success of the Seasat radar mission. However, from the present perspective, most were little more than digital efforts towards optical processing, and were cast in the same mold. They were spectral analysis machines designed within the conceptual constraints of optical SAR telescopes.

Paradigm challenged

Meanwhile, conceptually similar signal processing problems had been under study in other areas for many years. An important development was published by *Stolt* [1978] in the field of seismics. The challenge he addressed was to focus a set of signals from sub-surface echoes gathered by an extended string of geophones, a problem closely analogous to that of a synthetic aperture radar. Seismic data can have extreme range curvature, with nearly 180° angular (azimuth) field of view obtained in many situations. The predominant techniques then used in seismics were based on spectral analysis, and had been tuned to keep pace with digital processing capabilities. Stolt observed that the range curvature problem could be solved by an adroit change of variables in the two-dimensional frequency (wavenumber) domain [*Stolt*, 1978].

3.2 Wave Domain SAR Processing

Stolt's methods remained unknown to the SAR community until 1987, when an Italian mathematician and seismologist, while in residence at Stanford University, applied the Stolt change of variable to the SAR range curvature problem, with very good results. Fabio Rocca, together with colleagues, demonstrated that the wave domain technique, including a variation on the Stolt change of variables, could be used on Seasat SAR data [*Cafforio et al*, 1991]. Together with Richard Goldstein of the Jet Propulsion Laboratory, they applied the new technique to imaging radar interferometry. The technique provided much better geometric and phase fidelity in the complex image than did the conventional SAR processors then available.

Paradigm shift

The key to the wave domain curvature correction technique is that a phase multiply in the two-dimensional wave domain perfectly compensates for range curvature at the reference range chosen. The phase multiply trick is well known in Fourier transform operations used in spectral analysis, and has been established in radar analysis circles since the time of *Woodward* [1957], but it was a concept foreign to the optical SAR processing tradition.

The real lesson in the Stolt and Rocca approaches is that there are greater degrees of freedom in SAR processing strategy than an optical model allows, even for signal structure that has a similarity to optical imaging. Indeed, there is greater freedom than even a spectral analysis approach allows. It was only in 1989 or so that this realization had an impact on the field of synthetic aperture radar signal processing.

Stolt made his contribution by recognizing range curvature as the primary issue, not as a nuisance. His approach yielded results. Solution of the image focus problem followed as a direct consequence of solution of the range curvature problem. The new point of view caught on in the SAR community. A paradigm shift occurred.

The wavenumber domain

If data are transformed into the two-dimensional frequency domain, phase multiplies may be done simultaneously in the two dimensions. Processors have been built that do this, but there is a problem. In the wavenumber domain, the parameter of range is not available, yet all data have inherent range dependence for focus, for curvature, and for secondary range compression.

The Stolt change of variables is the complete transformation required to fully compensate for the oblique encounter geometry across the illuminated field. Calculation of the required expression is straight forward. It is another matter in a digital implementation, however. The Stolt change of variables [*Cafforio et al,* 1991], and variations on it [*Raney and Vachon,* 1989], have been applied to SAR processing, with positive results. A practical difficulty is that the Stolt change of variables requires an interpolator. This leads to increased computational burden. If processing speed is a major constraint, then it is tempting to implement the interpolator with a short kernal, thereby speeding up the process, but decreasing precision in the bargain. This often induces unacceptable phase errors in the complex image.

The chirp scaling algorithm

Most considerations of SAR and related imaging algorithms conceptualize the data in two domains, either in the frequency domain or in the time (or space) domain, both of which apply to either the range or the azimuth dimension of the data array. The logical departure necessary for understanding the chirp scaling algorithm [*Raney et al*, 1994] is to recognize the essential distinction between two aspects of the space representation: the image domain; and the signal domain. For the large time-bandwidth linear fm range pulse used by most Earth observing SARs, the signal domain in effect is a special transformation of the image domain. The chirp scaling algorithm takes advantage of the linear fm range pulse coding to fully correct range curvature, without interpolation [*Carrara, Goodman, and Majewski*, 1995].

For a linear fm (or chirp) waveform, conventional signal focusing is done using a matching conjugate chirp waveform. It is also known that if the chirp rate of the reference waveform is slightly incorrect, then the output pulse is shifted in position as well as being slightly out of focus. The resulting shift in the output pulse position is proportional to the differential fm chirp rate mis-match. Such a shift may be intentionally introduced in the processor and used to nullify the unwanted range curvature shift pattern of the input signals. This is the key to the chirp scaling algorithm.

The innovative step in this algorithm is to apply a special linear fm multiply to the data in the range signal / azimuth frequency domain. The fm rate of this phase multiply is chosen to exactly balance the range curvature difference between a signal at an arbitrary range and the curvature of a signal from the reference range. The chirp scaling phase multiply acts as a differential shift operator [*Papoulis*, 1968]. The total frequency excursion of the fm multiply is very small compared to the range bandwidth of the data so there are minimal unwanted consequences of its use.

Following this range chirp scaling phase multiply, a range Fourier transform leads to the two-dimensional frequency domain. As a result of the preceding step, all signals have the same curvature in the wavenumber domain. Therefore, phase multiplication can be applied to achieve range curvature correction as well as range pulse compression. Furthermore, secondary range compression can include a frequency dependent correction, which is a distinct improvement over the usual implementations. Note, in contrast to other methods of range curvature correction, that the required phase match in this domain is a single valued function which can be predicted accurately from the imaging geometry.

The remaining steps in the algorithm are relatively direct, even conventional. An inverse Fourier transform in range is needed to get to the range image / azimuth frequency domain. Here the usual azimuth matched phase is applied leading to azimuth compression, and an additional phase correction is required to pick up phase terms generated by the opening phase multiply. Azimuth inverse Fourier transformation leads to the final complex image.

No interpolation is needed in the chirp scaling algorithm. The complex image is generated through a sequence of range variant linear filters and complex multiplies. Using a reciprocal sequence of operations, the complex image derived through this algorithm could be inverted to replicate the original signal data set, range curvatures included. There would seem to be no other processing technique available that could pass such a test.

4 Fundamental Limits and Extensions

Most of the preceding discussion has focused on a conventional SAR system, i.e., a single channel radar that uses a relatively narrow real beamwidth oriented approximately orthogonal to the vehicle velocity vector. If one or more of these characteristics can be extended, then enhanced and/or special performance capabilities can be achieved.

Resolution may be improved in azimuth only through increased Doppler bandwidth. Bandwidth can be gained either by increasing the antenna beamwidth, or by increasing the angle over which a given region is illuminated. The latter is the basis for Spotlight SAR [*Walker*, 1980], in which high azimuth resolution is achieved at the expense of image size. For small range depth, the Stretch modulation technique [*Caputi*, 1971; *Walker*, 1980] is useful, which leads to improved range resolution without a large bandwidth penalty. Going the other direction, increased swath width may be obtained by trading resolution, which may be implemented using the ScanSAR technique.

With signals from two or more sets of mutually coherent channels, phase comparison between channels leads to rich new possibilities, including especially interferometry and polarization synthesis. Several specific interferometric schemes follow directly from the general case outlined below. If the observation channels are separated in the vertical plane, then interferometry leads to relative terrain height estimation. If the channels have a time delay between their otherwise identical observations, then very small movements in the illuminated scene may be measured. Together, the full mutual coherence of the scene can be estimated. An important and specialized use of the coherent structure between multiple channels of a SAR leads to quadrature polarimetry, through which the data from a SAR may be combined to replicate any combination of transmit and receive polarizations. The current state of the art in synthetic aperture radar is marked by continued developments in polarization synthesis and interferometry.

4.1 Resolution and Swath Width

In general, azimuth resolution is determined only by azimuth signal bandwidth, which, in turn, is determined by the radar wavelength λ and the span of angles (β_1, β_2) over which the object is viewed. The situation includes SAR, certain SONAR devices, Spotlight SAR, tomography, inverse SAR as used for imaging Earth satellite and nearby planets, etc. In all these cases, the azimuth (spatial) resolution r_A is given by the inverse (spatial) bandwidth, which leads to

$$r_A = \frac{\lambda}{2\left|\cos\beta_2 - \cos\beta_1\right|} \tag{1}$$

where the angles are measured with respect to the SAR velocity vector (neglecting Earth rotation).

A simple but effective form of aperture synthesis relies on rotation of the object, rather than motion of the radar past the object. Early experiments used a stationary radar observing test object on a rotating platform, for which specialized processing schemes were developed [*Walker*, 1980]. The technique is known as inverse SAR (sometimes abbreviated to ISAR). The method has been used with considerable success to image the moon and certain planets using large astronomical radars on Earth [*Evans and Hagfors*, 1968], artificial satellites in orbit around the Earth (which is also known as Space Object Identification, SOI), and to obtain "super-resolution" of certain moving objects, such as ships at sea [*Hardange*, 1992].

The minimum (limiting best, finest, or "highest") resolution is obtained when the effective field of view approaches 180°. Assuming that the object's reflectivity remains constant over the subtended angle, and that each scattering center is observable by the radar over this angle, then the limiting resolution is readily seen to be

$$\lim_{\substack{\beta_1 \to 90^\circ \\ \beta_2 \to -90^\circ}} r_A = \frac{\lambda}{4} \tag{2}$$

Although such spectacular performance may seem unrealistic, it should be noted that results approaching this value have been achieved in certain applications such as active seismics.

To image a swath much wider than ambiguity limits would normally allow, a SAR system can be designed to incorporate an alternative and less conventional technique [*Moore, et al.*, 1981; *Tomiyasu*, 1981]. This technique is now known as ScanSAR. RADARSAT is the first satellite SAR to use ScanSAR mode [*Luscombe*, 1988; *Raney et al.*, 1991]. RADARSAT was launched in November 1995, and has demonstrated the effectiveness of this technique.

The basis of ScanSAR is rapid steering of the elevation beam pattern of the antenna to build extended range coverage in steps. The resulting total swath width can be several times the range swath limit imposed by the ambiguity constraint on an individual beam. There is an associated cost, however: resolution of the resulting image is degraded in proportion to the increase in swath width.

ScanSAR relies on time-sharing between two or more separate subswaths. The set of returns used to image a section of one subswath must be from consecutive pulses to provide adequate sampling, and must be of sufficient length to allow formation of the synthetic aperture needed for the subswath at the required resolution. The imaging operations, therefore, are split up into a series of blocks of pulses, each block providing returns from one of the subswaths. Each block is processed to provide an image of a section of the corresponding subswath. The entire imaging sequence cycles around the full set of subswaths in a sufficiently rapid manner that the imaged sections in any one subswath are adjoining or overlapping.

As is true for any satellite SAR system, for a given pulse repetition frequency (prf) there are certain range positions which cannot be imaged because the timing of their reflections would coincide with a later pulse transmission. To obtain the very wide, continuous coverage that is the principal purpose of ScanSAR, it is necessary to use more than one prf across the set of subswaths.

Instead of relying on a continuous stream of data, as is normally done, the data from each subswath are received in the form of discrete blocks of echoes. This procedure is analogous to the burst imaging strategy used by the SAR on board *Magellan* when mapping Venus [*Johnson,* 1991]. Many of the algorithms developed for processing a continuous sequence of SAR data are neither appropriate nor efficient for burst mode operation. Additional processing requirements follow from the segmented form of the ScanSAR image. Radiometric and geometric matching are required in both along- and across-track directions to knit output data blocks into a continuous image file.

4.2 Interferometry

One could visualize a SAR as simply one example of a classical coherent imaging system. It is well known that such systems reach their pinnacle of quantitative performance when used in an interferometric mode, *i.e.*, one in which the scene observed is made to yield its secrets at scales of the wavelength of illumination. This is the objective of SAR configurations designed to use two coherent channels, separated either spatially or temporally.

A SAR image of terrain is a two-dimensional mapping of a three-dimensional surface. Unless more information is available, there is no way to quantitatively estimate terrain height with respect to a datum plane. Of course, as with aerial photographs, a pair of SAR images, each taken at different incident angles, could be used as a stereo pair. Being a coherent system, however, SAR has much more to offer than that.

The basic interferometric environment is one in which there is a pair of mutually coherent signals having the same envelope, carrier frequency, and phase modulation, but which have different reference phase. Signal processing is designed to estimate the phase difference, and to deduce geophysical parameters of interest from the resulting measurement. Even from a satellite's altitude, SAR interferometry is able to observe physical movements or other changes in the observed scene that are at the scale of the wavelength of the radar being used, typically, on the order of centimeters.

The interferometric signal model is simple in concept. For any neighborhood in the scene, let the complex image pair be described by

$$s_1(t) = \Gamma_{t_1} a \exp\{-j\varphi\} \tag{3}$$

$$s_2(t) = \Gamma_{t_2} a \exp\{-j\varphi + j\Delta\varphi(r, t_2 - t_1)\} \tag{4}$$

where the subscripts t_1 and t_2 suggest that the two signals may be obtained at two different points in time as well as from two different spatial perspectives. The objective is to estimate the relative phase difference $\Delta\varphi$. For this, there are two methods employed, correlation and sum-and-difference linear combination. Correlation, being the more generally applicable method, is based on the statistical norm

$$E[s_1(t_1)\, \bar{s}_2(t_2)] = R_{12}(t_1, t_2)\, a \exp\{-j\Delta\varphi(r, t_2 - t_1)\} \tag{5}$$

which is presented as it would be calculated using complex image data. Here, E[.] is the expectation (averaging) operator. Application of this formulation always depends on the cross-correlation $R_{12}(t_1,t_2)$ of the two scene functions as well as on the systematic phase difference between them. The phase difference $\Delta\varphi$ may be due either to geometric or to temporal differences between the two observations. Specifics of these possibilities are discussed further below.

Any approach to phase difference measurement is subject to the fundamental 2π ambiguity characteristic of phase estimation algorithms. In many radar situations, knowledge of the physical constraints of the situation coupled with phase unwrapping algorithms is sufficient for the purpose. More on the phase unwrapping problem may be found in [*Prati et al*, 1990].

Successful interferometry depends on the cross correlation $R_{12}(t_1,t_2)$ of the scattering functions Γ_{t1} and Γ_{t2}, which becomes by default an observable of the measurement. The normalized cross-correlation function, a useful form of the same statistic, is the mutual coherence function

$$\gamma_{12}(t_1, t_2) = \frac{R_{12}(t_1, t_2)}{\sqrt{\mathrm{E}\left[|s_1|^2\right]}\sqrt{\mathrm{E}\left[|s_2|^2\right]}} \tag{6}$$

frequently encountered in physical optics [*Born and Wolf*, 1959]. Use of either the cross-correlation function or the mutual coherence function allows quantitative representation of both spatial and temporal dependence of correlation loss between the two observations.

The three main classes of spatial interferometry are differentiated by conditions on the viewing geometry, by temporal properties of the signals used in the measurement, and by conditions on scene coherence. In the literature, SAR interferometry is known by various names, such as InSAR or ISAR. In each case, the form of interferometer intended must be clarified by context. In the following discussion, nomenclature is suggested that is meant to avoid ambiguity, even when taken out of context.

Range vector (delta-r) interferometry

A single channel radar can measure only the range delay and Doppler shift for each scattering object. In the three dimensional space of an observed scene, three parameters are required for scatterer location. Given knowledge of Doppler frequency, two more measurements are needed. Simple measurement of scalar range is not sufficient. Either range and scatterer altitude is needed, or else range and the elevation angle. In short, the range vector (range and direction) is required.

One method of measuring the range vector is for the radar to use two antennas rather than one [*Graham*, 1974; *Madsen et al.*, 1993]. This leads to the *delta-R* interferometric configuration. The two antennas in general both should be in a plane orthogonal to the motion vector of the radar, but they must be separated spatially in the elevation plane. The range varying phase difference between the two channels is found by forming the cross-correlation between the two received signals. In its simplest form, the two channels are operated simultaneously, so that the observed mutual coherence function of the scene is always unity. From any given transmission, the signals received at the two antennas have a phase difference determined uniquely by the geometry, and, of most interest, the combined range and elevation angle to each scattering element. Through interferometric signal analysis, a vector range record of reflectivity is obtained. This means that both range magnitude (distance) and range elevation angle (direction) are available for each mapped reflection. These numbers may be inverted to plot an orthographic terrain map, free of any terrain elevation distortion effects. Alternatively, the signal pair may be processed to obtain an estimate of the height h_i of each element of the scattering surface with respect to a datum plane. With this method of using one pair of signal sequences, obtainable elevation precision is on the order of the range resolution of the radar [*Prati et al*, 1989].

From geometric constraints, the spacing between the antennas should be on the order of one meter for aircraft radars. There are several such systems in operation. However, for orbital altitudes, the same configuration, in principle, would lead to an antenna separation on the order of a kilometer, which is not practicable for a single spacecraft. There is a way around the problem, however.

Double-pass delta-R interferometry

In the mid-1980s, a brilliant innovation was proposed by Richard Goldstein [*Zebker and Goldstein*, 1986] to circumvent the antenna spacing problem for satellite interferometric radars. Assuming adequate mutual scene coherence, spatial interferometry should be obtained from a pair of signals collected (and recorded) from one spacecraft operated so as to observe the scene from two different but neighboring positions, i.e., on different orbital passes. The technique has been proven to work with satellite SAR data [*Li and Goldstein*, 1990].

The radar passes must yield a pair of data sequences that emulate the geometry of simultaneous coverage. This implies two constraints: spatial, and temporal.

Assume at the outset that the scene reflectivity is the same for both passes of the radar, even at the detailed level of the radar wavelength. The spatial constraint applies to the spacing between the orbital passes. Ideally, the radar wavelength projected onto each area of the surface must be the same from both orbits. Since the two orbits are separated, each area is observed at a slightly different incident angle. This implies that the projected wavelength is slightly different for the two cases. Interference is supported only if the range bandwidth of the radar signal is sufficient to span the required wavelengths, which becomes more demanding as the orbit separation increases. Loss of mutual coherence through increasing orbit spacing is known as baseline decorrelation [*Gabriel and Goldstein, 1988*]. For typical large time-bandwidth product signals, the elevation angle constraint implies that correlation of the pair of return signals is maintained for orbital separations no larger than a few kilometers. For absolute elevation maps to be derived, however, precise knowledge is required of the spatial separation between observation tracks.

The second consideration has to do with scene coherence. In order for the two signals to act as an interferometric pair, their respective phase structures must be robust over the time interval between satellite observations. In short, there must exist mutual coherence between the two images, even though they each derive from different observation times. This requirement is more easily satisfied for short inter-opportunity intervals, such as from the three day repeat orbit first used with Seasat data to prove the concept, and for stable terrain features, such as unvegetated rocky mountain slopes. It is not necessarily satisfied for scenes, such as ice, that may undergo changes in the details of reflection and scattering between observations. For example, in one unpublished experiment, elevation contours were derived from two-pass ERS-1 SAR data for a mountainous region of western Canada. The results were satisfactory, except for a glaciated slope for which the value of the mutual coherence function approached zero. Evidently, the glacier's reflectivity had changed between observations.

Differential delta-R interferometry

Consider two pairs of delta-R interferometric measurements which share a common reference phase, φ_0. The necessary conditions for interference may be satisfied by using data from three single passes, assuming suitably stable scene correlation, or by two passes of a two-channel delta-R interferometer. In either case, the differential phase measurement

$$\Delta(\Delta\varphi) = (\varphi_1 - \varphi_0) - (\varphi_2 - \varphi_0) = \varphi_1 - \varphi_2 \tag{7}$$

exploits the common reference phase φ_0 as a transfer standard of range/elevation. Thanks to this phase reference, the interferometric measurement has a precision that is on the order of the illuminating wavelength [*Prati et al.*, 1989]. The technique has been demonstrated and used to indicate displacement on the order of centimeters between observations of objects [*Gray and Farris-Manning*, 1993] and of distributed terrain features [*Gabriel et al.*, 1989].

Time difference (delta-t) interferometry

A systematic phase difference between two otherwise identical signals will occur if they are gathered at different times, between which there is movement in the observed scene. A delta-t, or time difference, interferometer is based on the use of two antennas separated by distance D along the motion vector of the radar having along track velocity V, leading to a difference δt in observation time of the scene. If there are elements of the scene with velocity v_{rad} towards the radar, then a phase difference $\Delta\varphi = 2k\,\delta t\,v_{rad}$ will occur. (The time delay δt between the pair of received signals is D/V

if the antennas are operated separately as transmitter and receiver. The time delay is *D/2V* if, as is usually the case, only one antenna is used for transmission and both are used for reception.) The phase shift may be used to detect small radial movement of scatterers from a moving platform [*Urkowitz*, 1964], a principle that applies even in a high-resolution SAR context [*Raney*, 1971]. Delta-t interferometry has been used successfully to measure ocean surface currents [*Goldstein and Zebker*, 1987].

Just as for other forms of interferometry, there must be mutual coherence of the respective scattering functions in order to successfully estimate their phase difference. The mutual coherence function itself can be used as an image interpretation tool, mapping areas of different coherence levels within an area of interest.

4.3 Polarization

Polarization describes the spatial orientation of the *E* vector of an electromagnetic wave. Most radars operate with a single antenna, which not only determines the polarization of the transmitted wave, but also receives the same polarization component of the backscattered waves. Such a radar may be viewed as providing only a scalar estimate of the scene's reflectivity for a given polarization, which has been the basis for much of the preceding discussion.

To capture the complete polarization characteristics of either the transmitted em wave or the received wave, an electromagnetic vector description is required. For any given polarization of the transmitted wave, the reflectivity process in general will give rise to a diversity of polarizations in the backscattered wave. In most situations, additional information about the desired scene features may be deduced from the vector transformation measurement embedded in the received data. To observe these, the radar must be built such that more than one polarization is capable of being received. Similarly, the reflectivity is a function of the polarization of the transmitted wave. Thus, more than one polarization must be transmitted as well. Implementation of these features requires more transmitted power and greater data capacity, as well as an antenna able to transmit and receive more than one polarization.

There are several versions of polarization-diversity radars. The simplest form consists of two differently polarized antennas and their attendant receivers and recorders. Data from the two channels may be compared through their (power) ratio. The most elegant form, known as quadrature-polarimetry, consists of compiling a data set sufficiently complete so that the reflectivity of each area of the scene may be analyzed over all possible combinations of transmit and receive polarizations. Although perfect channel isolation is not possible in practice, calibration procedures are sufficient to compensate for realistic levels of inter-channel contamination. It is customary in radar remote sensing to designate the horizontal (H) and the vertical (V) orientations as the orthogonal polarization basis vectors.

Mutually incoherent channels

If the relative phase between polarization channels is not known, then phase cannot be used for polarimetric processing and interpretation. In this case, polarization diversity is restricted to only the mean reflected power corresponding to each pair of radar polarization states. Consider a radar that uses (V, H) polarizations on both transmit and receive. An estimate of the reflectivity coefficient σ^0 is obtained at every point in the image for all four combinations of transmit and receive orientations, leading to

$$\begin{bmatrix} \sigma^0_{HH} & \sigma^0_{HV} \\ \sigma^0_{VH} & \sigma^0_{VV} \end{bmatrix} \tag{8}$$

as the data available for subsequent processing. In this notation, the first subscript is taken to mean the polarization state of the received wave, and the second subscript describes the state of the transmitted wave. Although this order of subscripting is the standard in recent radar polarimetric literature, it is by no means standard for all of the radar (or optical) literature of relevance to the field. This matrix is sometimes referred to as the "scattering matrix", although more correctly it is the sigma nought matrix. The term scattering matrix properly refers only to the array of complex reflectivity coefficients.

There are several additional observations that might be made at this point. First, as is satisfied in most situations by almost all radars, reciprocity applies, such that the transmit and receive conditions may be interchanged, whereby $\sigma^0_{HV} = \sigma^0_{VH}$. Thus, there are only three independent estimates of power reflectivity for each scene element.

Second, transformation of the transmitted polarization state by reflection requires two or more actual reflections within the scene before the wave assumes its backscattered direction. Upon each reflection, the wave amplitude is reduced. Except for very special circumstances, therefore, the cross-polarized component always is weaker than the like-polarized component. On average, the difference is about 6 dB when observing linearly polarized signals, with a range of 2 dB to 10 dB commonly encountered [*Long*, 1975]. If one is interested in the cross-polarized signal in applications, then the polarization loss must be included in the relevant signal-to-noise ratio or other form of the radar equation during system design stages. Substantially more transmitter power is needed than would be required otherwise. This is one reason that most remote sensing radars must be satisfied with only one channel of like polarization.

Third, the strength of reflectivity in one polarization state relative to another polarization state often is a useful parameter. This ratio is determined primarily by the physical characteristics of the reflecting element, and rather little by radar system characteristics such as end-to-end power gain. Knowledge of the relative gain between channels is sufficient. In this case, the co-polarization ratio (e.g., σ_{HH}/σ_{VV}) and the depolarization ratio (e.g., σ_{VH}/σ_{VV}) have quantitative significance.

Fourth, as an illustration of the incompleteness of the incoherent representation, note the following. Although H and V are orthogonal, the magnitude of the reflectivity with respect to these two orientations is not sufficient to allow transformation into any other polarization state. For example, it is not possible to transform $[\sigma_{HH}\ \sigma_{VH}\ \sigma_{VV}]$ data into an estimate of the reflectivity matrix of a scene in terms of circular polarization components. To do so requires phase information.

Quadrature polarization

When four mutually coherent channels having suitable orthogonal polarizations (two on transmission and two on reception) are used to observe the scene, the radar is said to be quadrature-polarimetric, or "quad-pol" for short. Note that quad-pol radars are to be distinguished from less elegant polarization diversity radars. With a coherent radar, by definition, phase is available in the complex image for subsequent processing. A powerful principle comes to bear at this point. When both phase and amplitude information are available for the elements of the scattering matrix, then transformation from one polarization basis to any other may be accomplished by a simple matrix multiply [*Nespor et al.*, 1985; *Giuli*, 1986]. Similarly, in the Stokes matrix representation, coherent observation of scene reflectivity for one polarization state provides sufficient data for transformation to any other polarization combination.

Polarimetric synthesis requires several steps. For each pixel, all elements of the Stokes matrix must be found. These are calculated from the recorded data files by using the known orthogonal transmit and receive Stokes vectors. Once the matrix elements are known, then the resulting matrix may be used to transform a vector description of a transmitted field into a vector description of the reflected field. In turn, this is selectively observed through the receive vector, from which the image brightness distribution may be calculated. The approach requires that the original data be relatively free of contamination from cross-talk between polarization channels. This implies stringent requirements for channel isolation (hardware) and system calibration (procedure).

It follows that a quad-pol data set may be analyzed over all states to determine the polarization of reflection from each neighborhood in a scene. The resulting plot is known as a polarization signature [*van Zyl,* 1989], which exposes the relative intensity of the like-polarized and the cross-polarized components as a function of the angular parameters of the polarization ellipse. These signatures may be compared to theoretical predictions for certain classes of reflectivity, such as Bragg scattering. This approach offers a powerful way to determine the dominant reflectivity mechanisms in the scene. For example, the reflectivity from the canopy of a tropical forest may be separated from reflectivity components arising from the two-bounce path of tree trunks emerging from standing water. This and related applications are nicely summarized by [*Zebker et al,* 1991].

Most quad-pol data available to the community have been derived through AirSAR of JPL [*van Zyl et al,* 1991; *Way and Smith,* 1991], and the SIR-C/X-SAR mission [*TGARS,* 1995]. The JPL data are in compressed Stokes matrix format. In this context, "compressed" implies two things [*Zebker et al.,* 1987; *van Zyl et al.,* 1991] : dynamic range; and averaging. Data dynamic range is reduced through control of the number of bits used in the matrix, and by scaling with respect to total power. Averaging is done over the corresponding matrix elements that belong to four adjacent azimuth image pixels. This averaging, which is equivalent to multi-looking, is possible because the Stokes matrix elements are proportional to power.

5 Closing Comments

Three aspects of synthetic aperture radar have been addressed: SAR history, where the evolution of processing provides a penetrating perspective; SAR frontiers, covering selected fundamental limits as well as current capabilities especially interferometry and polarimetry; and SAR theory, encapsulated for the first time in quasi-axiomatic principles. Hopefully, the discussion has succeeded in explaining and defending SAR as a relatively mature, precise, and robust technique. Apologies are extended to those who may harbor disappointment or annoyance with the unconventional style employed by the author for this article.

References

Bamler, R. [1992], A Comparison of Range-Doppler and Wavenumber Domain SAR Focusing Algorithms, *IEEE Transactions on Geoscience and Remote Sensing,* vol. 30, no. 4, pp 706-713.

Bennett, J. R., I. G. Cumming, and R. A. Deane [1980], The Digital Processing of

Seasat Synthetic Aperture Radar Data, *Proceedings of the IEEE International Radar Conference*, pp 168-174.

Bennett, J. R. and P. R. McConnell [1980], Considerations in the Design of Optimal-Multi-look Processors for Image Quality, *Proceedings of the 3rd Seasat SAR Workshop*, Frascatti, ESA-SP-172, December, pp 5-12.

Born, M. and E. Wolf [1959], *Principles of Optics*, Pergamon Press, MacMillan, New York.

Brown, W. M. [1967], Syntheic Aperture Radar, *IEEE Transactions on Aerospace Electronic Systems*, AES-3, pp 217-229.

Brown, W. M. and R. J. Fredricks [1969], Range-Doppler Imaging with Motion Through Resolution Cells, *IEEE Transactions on Aerospace Electronic Systems*, AES-5, pp 98-106.

Cafforio, C., C. Prati, and F. Rocca [1991], Full Resolution Focusing of Seasat SAR Images in the Frequency-Wavenumber Domain, *International Journal of Remote Sensing*, vol 12, pp 491-510.

Caputi, W. [1971], Stretch: A Time-Transformation Technique, *IEEE Transactions on Aerospace and Electronic Systems*, AES-7, no.2, March, pp 269-278.

Cutrona, L. J. and G. O. Hall [1962], A Comparison of Techniques for Achieving Fine Azimuth Resolution, *IRE Transactions on Military Electronics*, MIL-6, pp 119-133.

Carrara, W. G., R. S. Goodman, and R. M. Majewski [1995], *Spotlight Synthetic Aperture Radar: Signal Processing Algorithms*, Artech House, Boston, 1995.

Cutrona, L. J., E. N. Leith, C. J. Palermo, and L. J. Porcello [1960], Optical Data Processing and Filtering Systems, *IRE Transactions on Information Theory*, vol. IT-6, pp-386-400.

Cutrona, L. J., E. N. Leith, L. J. Porcello, and W. E. Vivian [1966], On the Application of Coherent Optical Processing Techniques to Synthetic-Aperture Radar, *Proceedings of the IEEE*, vol. 51, no. 8, pp 1026-1032.

Evans, J. V. and T. Hagfors [1968], *Radar Astronomy*, McGraw-Hill Book Company, New-York

Gabriel, A. K. and R. M. Goldstein [1988], Crossed Orbit Interferometry: Theory and Experimental Results from SIR-B, *International Journal of Remote Sensing*, vol. 9, no. 5, pp 857-872.

Gabriel, A. K., R. M. Goldstein, and H. A. Zebker [1989], Mapping Small Elevation Changes over Large Areas: Differential Radar Interferometry, *Journal of Geophysical Research*, vol. 94, no. B7, July, pp 9183-9191.

Giuli, D. [1986], Polarization Diversity in Radars, *Proceedings of the IEEE*, vol. 74, no. 2, February, pp 245-269.

Goodman, J. W. [1976], Some Fundamental Properties of Speckle, *Journal of the Optical Society of America*, vol. 66, no. 11, pp 1145-1150.

Gray, A. L. and P. Farris-Manning [1993], Repeat-Pass Interferometry with Airborne Synthetic Aperture Radar, *IEEE Transactions on Geoscience and Remote Sensing*, vol. 31, no. 1, pp 180-191.

Hardange, J. P. [1992], Inverse Synthetic Aperture Radar, Chapter 9, *Fundamentals and Special Problems of Synthetic Aperture Radar*, Lecture Series 182, AGARD, NATO, Paris.

Harger, R. O. [1970], *Synthetic Aperture Radar Systems: Theory and Design*, Academic Press, New York.

Jin, M. Y. and C. Wu [1984], A SAR Correlation Algorithm which Accommodates Large Range Migration, *IEEE Transactions on Geoscience and Remote Sensing*, vol. 22, pp 592-597.

Johnson, W. T. K. [1991], Magellan Imaging Radar Mission to Venus, *Proceedings of the IEEE*, vol. 79, no. 6, pp 777-790.

Kozma, A. E. N. Leith, A. VanderLugt, and W. Hall [1972], Tilted-Plane Optical Processor, *Applied Optics*, vol. 11, August, pp 1766-1777.

Kuhn, T. H. [1970], *The Structure of Scientific Revolutions*, Second Edition, The University of Chicago Press, Chicago.

Leith, E. N. [1971], Quasi-Holographic Techniques in the Microwave Region, *IEEE Spectrum*, vol. 59, Sept., pp 1305-1318.

Leith, E. N. [1973], Range-Azimuth Coupling Aberrations in Pulse-Scanned Imaging Systems, *Journal of the Optical Society of America*, vol. 63, pp 119-126.

Leith, E. N., and J. Upatnieks [1965], Reconstructed Wavefronts and Communication Theory, *Journal of the Optical Society of America*, vol. 52, pp 1123-1130.

Li, F. L. and R. M. Goldstein [1990], Studies of Multibaseline Spaceborne Interferometric Synthetic Aperture Radars, *IEEE Transactions on Geoscience and Remote Sensing*, vol. 28, no. 1, pp 88-97.

Long, M. W. [1975], *Radar Reflectivity of Land and Sea*, Lexington Books, D. C. Heath, Toronto/London.

Luscombe, A. P. [1988], Taking a Broader View: RADARSAT Adds ScanSAR to its Operation, *Proceedings of the International Geoscience and Remote Sensing Symposium IGARSS '90*, Edinburgh, Scotland, pp 1027-1032.

Madsen, S. N., H. A. Zebker, and J. Martin [1993], Topographic Mapping Using Radar Interferometry: Processing Techniques, *IEEE Transactions on Geoscience and Remote Sensing*, vol. 31, no. 1, pp 246-256.

Moore, R. K. [1979], Trade-off Between Picture Element Dimensions and Noncoherent Averaging in Side-looking Airborne Radar, *IEEE Transactions on Aerospace and Electronic Systems*, vol. 15, Sept., pp 697-708.

Moore, R. K., J. P. Claasen, and Y. H. Lin [1981], Scanning Spaceborne Synthetic Aperture Radar with Integrated Radiometer, *IEEE Transactions on Aerospace and Electronic Systems*, AES-17, no. 3, pp 410-421.

Nespor, J. D., A. P. Agrawal, and W. M. Boerner [1985], Theory and Design of a Dual Polarization Radar for Clutter Analysis, *Inverse Methods in Electromagnetic Imaging*, W. M. Boerner et al., Eds., Hingham, MA, Reidel, pt. 1, pp 643-659.

Papoulis, A. [1968], *Systems and Transforms with Applications to Optics*, McGraw-Hill, New-York.

Pettengill, G. H. [1993], Radio and Radar Exploration of Venus: Highlights of the Magellan Mission, *Modern Radio Science*, Hiroshi Matsumoto, Editor, Oxford University Press, pp. 29-52.

Prati, C., M. Giani, and N. Leuratti [1990], SAR Interferometry, A 2D Phase Unwrapping Technique Based on Phase and Absolute Value Information, *Proceedings of the International Geoscience and Remote Sensing Symposium, IGARSS '90*, Washington, D.C., pp 2043-2046.

Prati, C., F. Rocca, and A. M. Guarnieri [1989], Effects of Speckle and Additive Noise on the Altimetric Resolution of Interferometric SAR (ISAR) Surveys, *Proceedings of the International Geoscience and Remote Sensing Symposium, IGARSS '89*, Vancouver, Canada, pp 2469-2472.

Raney, R. K. [1968], *Theory of Quadratic Systems*, Ph.D. Thesis, University of Michigan, University Microfilms, Ann Arbor, Michigan.

Raney, R. K. [1971], Synthetic Aperture Radar and Moving Targets, *IEEE Transactions on Aerospace and Electronic Systems*, AES-7, no. 3, pp 499-505.

Raney, R. K. [1985], Theory and Measure of Certain Image Norms in SAR, *IEEE Transactions on Geoscience and Remote Sensing*, vol. 23, no. 3, pp 343-348.

Raney, R. K. and P. W. Vachon [1989], A Phase Preserving SAR Processor, *Proceedings of the International Geoscience and Remote Sensing Symposium, IGARSS '89*, Vancouver, Canada, pp 2588-2591.

Raney, R. K., A. P. Luscombe, E. J. Langham, and S. Ahmed [1991], RADARSAT, *Proceedings of the IEEE*, vol. 79, no. 6, pp 839-849.

Raney, R. K., H. Runge, R. Bamler, I. Cumming, and F. Wong [1994], Precision SAR Processing Using Chirp Scaling, *IEEE Transactions on Geoscience and Remote Sensing*, vol. 32, no. 4, pp. 786-799.

Sherwin, C. W., J. P. Ruina, and R. D. Rawcliffe [1962], Some Early Developments in Synthetic Aperture Radar Systems, *IRE Transactions on Military Electronics*, MIL-6, pp-111-115.

Stolt, R. [1978], Migration by Fourier Transform Techniques, *Geophysics*, vol. 43, no. 1, pp 49-76.

TGARS [1995], *IEEE Transactions on Geoscience and Remote Sensing, Special Issue on SIR-C/X-SAR*, vol. 33, no. 4.

Tomiyasu, K. [1981], Conceptual Performance of a Satellite Borne, Wide Swath Synthetic Aperture Radar, *IEEE Transactions on Geoscience and Remote Sensing*, vol. 19, no. 2, pp 108-116.

Urkowitz, H. [1964], The Effect of Antenna Pattern on the Performance of Dual-antenna Radar Airborne Moving Target Indicators, *IEEE Transactions on Aerospace and Naval Electronics*, ANE-11, pp 218-223.

Van Zyl, J. J. [1989], Unsupervised Classification of Scattering Behavior using Radar Polarimetry Data, *IEEE Transactions on Geoscience and Remote Sensing*, vol. 27, no. 1, pp 36-45.

Van Zyl, J. J. et al. [1991], *AIRSAR Reference Manual*, Jet Propulsion Laboratory, California Institute of Technology, Pasadena, California.

Walker, J. [1980], Range-Doppler Imaging of Rotating Objects, *IEEE Transactions on Aerospace and Electronic Systems*, AES-16, no. 1, pp 23-52.

Way, J. and E. A. Smith [1991], The Evolution of Synthetic Aperture Radar Systems and their Progression to the EOS SAR, *IEEE Transactions on Geoscience and Remote Sensing*, vol. 29, no. 6, pp 962-985.

Wiley, C. A. [1985], Synthetic Aperture Radars - A Paradigm for Technology Evolution, *IEEE Transactions on Aerospace Electronic Systems*, AES-21, pp 440-443.

Woodward, P. M. [1957], *Probability and Information Theory, with Applications to Radar*, Pergamon Press, MacMillan, New York.

Wu, C. [1980], A Digital Fast Correlation Approach to Produce Seasat SAR Imagery, Proc. *IEEE International Radar Conference*, pp 153-160.

Zebker, H. A. and R. M. Goldstein [1986], Topographic Mapping from Interferometric Synthetic Aperture Radar Observations, *Journal of Geophysical Research*, vol. 91, no. B5, pp 4993-4999.

Zebker, H. A., J. J. van Zyl, S. L. Durden, and L. Norikane [1991], Calibrated Imaging Radar Polarimetry: Technique, Examples, and Applications, *IEEE Transactions on Geoscience and Remote Sensing*, vol. 29, no. 6, pp 942-961.

Zebker, H. A., J. J. Van Zyl, and D. N. Held [1987], Imaging Radar Polarimetry from Wave Synthesis, *Journal of Geophysical Research*, vol. 92, no. B1, pp 683-701.

NONLINEAR PHYSICS AND CHAOS

Werner Lauterborn and Ulrich Parlitz

1 Introduction

To be sure, our description of nature is far from being complete. However, we are improving constantly in many directions, one of them being nonlinear physics. Indeed, we have learned that nonlinearity introduces a wealth of genuine phenomena. Among them, chaotic dynamics is the most puzzling and intriguing one. This type of irregular dynamics calls for new methods to cope with and has already set out for applications. The investigations on chaotic dynamics have shed new light on the notions of predictability and determinism - and their relation -, quite fundamental issues. The deterministic dynamics of even simple systems may be notoriously difficult to predict and even impossible in the long run. Such systems abound in nature, and examples can be drawn from almost any scientific discipline (see, for instance, [1-7]). This tutorial and review presents an overview of the achievements and some present research activities in the field of nonlinear dynamics. In particular, questions of embedding data into high-dimensional spaces, of modelling and prediction, noise reduction, detecting determinism in time series, and secure communication via chaotic synchronization are addressed.

2 Basic notions

In this section the notions of nonlinear dynamics are briefly reviewed [8] to set the stage for recent achievements [9].

2.1 State space

The state of a dynamical system is most intuitively visualized geometrically as a point in a space, called *state space* or phase space, spanned by all the dependent variables of the system. Variables may be voltages, currents, temperatures, pressures, etc. In Fig. 1 these variables are called x_1, x_2, x_3, ...x_m,The initial state, at time t_0 of the hypothetical system is marked as x_0. The dynamic evolution, i. e. the change of the system with time, then is given by a curve in the state space of the system, called *trajectory* or *orbit.* Time is a parameter to be affixed to the points of the curve, as shown by the two times t_1 and t_2. At these times the system is characterized by the values of the variables given by the respective points they are affixed, to. Alternatively, the time may be added to the list of dependent variables as x_{m+1} to yield an extended state space.

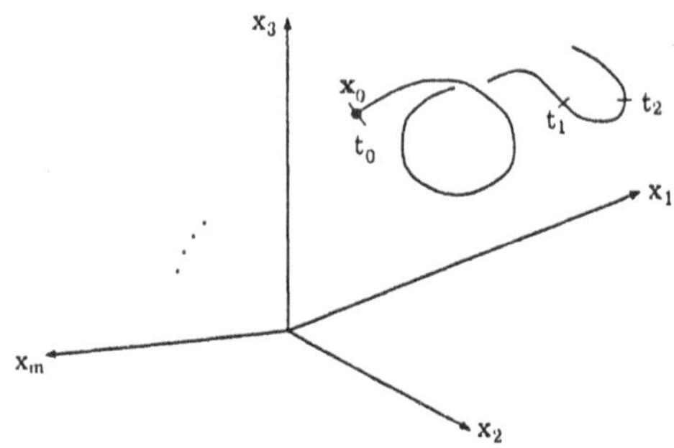

Figure 1: State space and a trajectory.

2.2 Attractors

It is a simple but intriguing question what happens to a whole volume of initial conditions as time proceeds. The gross answer is given by the theorem of Liouville. The volume stays constant in conservative systems although it may deform in a most peculiar way, and it shrinks in dissipative systems as they are usually encountered in physics. It is clear that the rate of shrinking depends on the physical system (or its model) as does the final volume (the limit set) and its form. The limit set for $t \rightarrow \infty$ is composed of one or more *attractors.* Each attractor has its own set of initial conditions leading to it in the course of time, the set being called its *basin.* There exist different types of attractors. The most simple one is a *fixed point,* i. e., a single permanent state—an equilibrium or state of rest, where there is no alteration anymore.

The next, more involved attractor, is called a *limit cycle.* It is a closed curve in state space that is traversed again and again by the system. Any self-excited oscillator, for instance the heart beat, belongs to this class. Radio and TV (still) rely on limit cycles (simple sine waves in this case) as carriers of information that is modulated onto them. A fixed point, in whatever high-dimensional state space, is zero-dimensional. A limit cycle is one-dimensional.

The next simple attractor fills a two-dimensional manifold, a *torus*. A trajectory on the torus is a *quasiperiodic* motion. The quasiperiodic motion on a torus comes about through two incommensurable frequencies being present simultaneously.

The most intriguing and variable type of attractor is the *strange* or *chaotic* attractor. It needs an at least three-dimensional state space to occur in the case of continuous dynamics and has a *fractal dimension*, not an integer one (0, 1, 2 in the previous cases), except for very special, accidental cases. Figure 2 shows graphical representations of the four types of attractors mentioned. The abundance of strange attractors in all disciplines of science now is a well established fact (see, e.g., [1-6]). Their characterization is mainly done via their fractal dimension (static property) and their Lyapunov spectrum (dynamic property). These notions are discussed below after having revealed how strange attractors may be obtained from measured data.

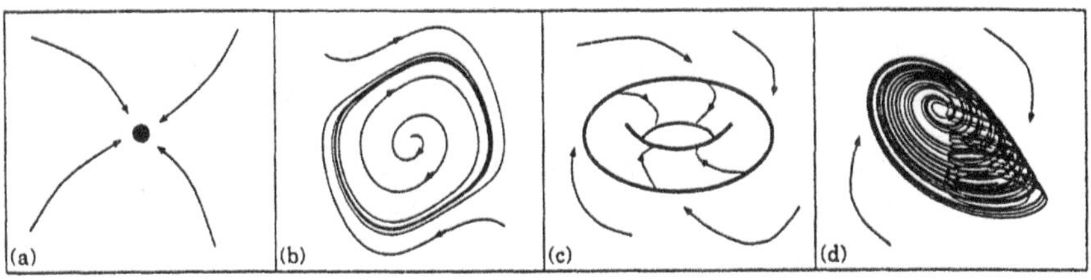

Figure 2: Types of attractors: (a) fixed point, (b) limit cycle, (c) torus, (d) projection of a strange attractor.

2.3 Poincaré sections

As fractal attractors of continuous dynamical systems must be more than two-dimensional, they can be visualized in printing (plane of the paper) only by projection. This often yields unwieldy pictures except for the beautiful standard examples of the Lorenz or the Rossler attractor (Fig. 2d). Often a better view of the structure of the attractor can be obtained by a method invented by Poincaré, whereby one dimension can be gained. It develops its full power for attractors in three-dimensional state spaces. In this method for visualizing an attractor a plane S, the Poincaré section plane (a hyperplane of dimension m -1 in general), is chosen in the state space that is intersected transversally by the trajectories. To display an attractor only the section points of the attractor in the plane S are plotted. The points hop around in the plane. For a quasiperiodic motion they group themselves to a closed curve (dimension one). They form definite structures with (presumably) selfsimilar properties in the case of a strange attractor and show no signs of structure in the case of white (or also coloured) noise. Figure 3 shows an example of a strange attractor of the driven, double-well Duffing oscillator $\ddot{x} + d\dot{x} - x + x^3 = a \sin \omega t$ for $d = 0.2$, $a = 0.3$, and $\omega = 1.24$. The Poincaré section plane is chosen as $0 = \omega t \bmod 2\pi$ in the extended state space. The folded structure of the attractor, not at all consisting of homogeneously scattered points, can be noticed. As the times at which the system is sampled are successive integer multiples of the period of the driving, the method in this case just delivers a stroboscopic phase portrait of the system as well known to physicists and engineers.

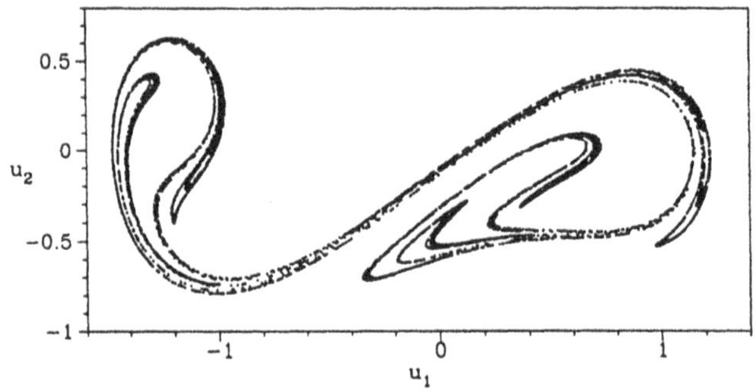

Figure 3: A strange attractor of the double-well Duffing oscillator $\ddot{x} + dx - x + x^3 = a \sin \omega t$ *for* $d = 0.2$, $a = 0.3$, *and* $\omega = 1.24$ *in a Poincaré section plane.*

2.4 Bifurcations

A system normally depends on some parameters, e. g., temperature, and may show a totally different behaviour - expressing itself in a totally different attractor - at a different parameter value. For instance, a fixed point may change into a limit cycle signalling that the system now has acquired the ability to oscillate. Such a change in the qualitative behaviour of a system is called a *bifurcation*. A bifurcation occurs at some parameter value called a *bifurcation point*. It is a point in *parameter space* that may be high-dimensional when there are many parameters. A system or, more precisely, a family of systems then is best characterized by the set of parameter values at which a bifurcation occurs, the *bifurcation set*. When the bifurcation set is plotted in the parameter space, it generates regions of topological similar behaviour and thereby a *parameter space diagram*, also called *phase diagram*.

There are four basic types of local bifurcations, the *period-doubling*, the *saddle-node*, the *transcritical*, and the *Hopf* bifurcation [10]. The period-doubling bifurcation has become widely known through its scaling behavior [11]. This bifurcation likes to appear in a cascade when a parameter is altered (Fig. 4), and often there appears a complete, i. e., infinite cascade of period doublings leading to aperiodic motion via a period that becomes infinitely long.

A saddle-node bifurcation appears, e.g., at those points, where the resonance curve of a nonlinear oscillator turns over. At these points, the oscillation gets unstable, and the system has to settle down somewhere else.

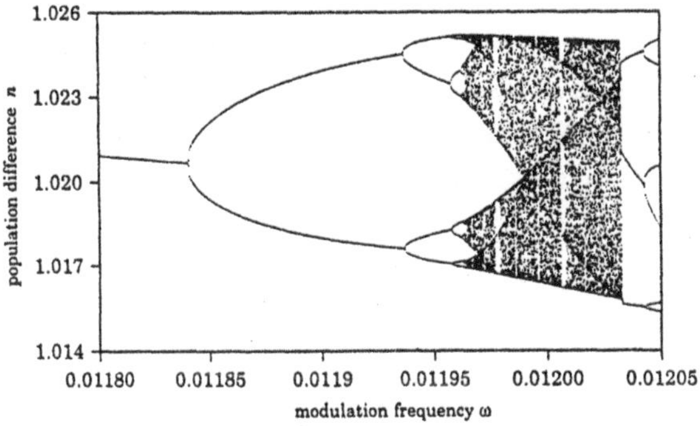

Figure 4: Bifurcation diagram of the periodically pumped ideal four-level laser:

$$\dot{q} = -q + nq + sn, \quad \dot{n} = p_0(1 + p^m \sin \omega t) - bn - nq$$

for $s = 10^{-7}$, $p_0 = 6 \cdot 10^{-4}$, $p_m = 1$, *and* $b = 0$ *(from Ref. [12]).*

The transcritical bifurcation appears in systems, where coexistent stable and unstable fixed points exist and where one of them does not depend on a parameter. Then, upon alteration of this parameter, a stable and an unstable fixed point may collide and exchange their stability at a bifurcation point via a transcritical bifurcation. The simple laser rate equations (see, e.g., [12]) without spontaneous emission term show a transcritical bifurcation for the photon number at the laser threshold.

At a Hopf bifurcation a limit cycle is generated. Self-excited oscillations derive their limit cycle from a Hopf bifurcation. The standard example is the van der Pol oscillator ([13], see also Fig. 2).

3 Nonlinear time series analysis

In mathematical models of dynamical systems the dynamics is visualized in their state space, whose (integer) dimension is given by the number of the dependent variables of the model. In experiments, usually just one variable is measured as a function of time, and the state space is not normally known. How, then, to arrive at the attractor that may characterize the system? This gap between the theoretical notions and observable quantities was filled in 1980 when *Packard, Crutchfield, Farmer,* and *Shaw* [14] published their fundamental paper "Geometry from a time series" where for the first time state space reconstruction methods were applied to scalar time series. A mathematical justification of this approach was given by *Takens* [15] at about the same time. He proved that it is possible to (re)construct, from a scalar time series only, a new state space that is diffeomorphically equivalent to the (in general unknown) original state space of the experimental system.

Essentially two methods for reconstructing the state space are available: *delay coordinates* and *derivative coordinates*. Derivative coordinates were used by *Packard* et *al.* [14] and consist of higher order derivatives of the measured time series. Since derivatives are susceptible to noise, derivative coordinates usually are not very useful for experimental data. Therefore, we will discuss the method of delay coordinates only.

3.1 Embedding

Let M be a smooth (C^2) m-dimensional manifold that constitutes the (original) state space of the dynamical system under inverstigation and let $\phi^t : \mathrm{M} \to \mathrm{M}$ be the corresponding flow (see [10] for a definition). Suppose that we can measure some scalar quantity $s(t) = \mathrm{h}(\mathbf{x}(t))$ that is given by the measurement function $\mathrm{h}: \mathrm{M} \to \mathbb{R}$, where $\mathbf{x}(t) = \phi^t(\mathrm{x}(0))$. Then one may construct a *delay coordinates map*

$$\begin{aligned} &F: M \to \mathbb{R}^d \\ &\mathbf{x} \mapsto \mathbf{y} = \mathrm{F}(\mathbf{x}) = (s(t), s(t+t_l), s(t+2t_l), \ldots, s(t+(d-1)t_l)) \end{aligned} \tag{1}$$

that maps a state x from the original state space M to a point y in a reconstructed state space $\mathbb{R}^d$, where d is the *embedding dimension* and t_l gives the *delay time* (or *lag*) used. *Takens* [15] proved that for $d \geq 2m + 1$ it is a generic property of F to be an embedding of M in $\mathbb{R}^d$, i.e., $F: M \to F(M) \subset \mathbb{R}^d$ is a (C^2–) diffeomorphism. Generic means that the subset of pairs (h, t_l) which yield an embedding is an open and dense subset in the set of all pairs (h, t_l). This theorem was generalized by *Sauer, Yorke and Casdagli* [16, 17] in two ways. Firstly, they replaced the condition $d \geq 2m + 1$ by $d > 2d_0(A)$ where $d_0(A)$ denotes the capacity (or: box-counting) dimension of the attractor $A \subset M$. This is a great progress for experimental systems that possess a low-dimensional attractor (e.g., $d_0(A) < 5$) in a very high-dimensional space (e.g., $m = 100$). In this case, the theorem of *Takens* says that only for very large embedding dimensions d (e.g., $d \geq 201$) we can expect a diffeomorphic equivalence for sure, whereas with the condition of *Sauer et al.* a much smaller d will suffice (e.g., $d > 10$). The second improvement of *Takens'* embedding theorem by *Sauer et al.* refers to the term "generic" that was replaced by "prevalent" which essentially means "almost all (h, t_l)" will yield an embedding. This modification of the original theorem was necessary, because examples of open and dense (=generic) sets have been found that are rather "thin" [16, 17]. Furthermore, *Sauer et al.* showed that for dimension estimation an embedding dimension $d > d_0(\mathrm{A})$

suffices. In this case the delay coordinates map F is, in general, not one-to-one, but the points where trajectories intersect are negligible for dimension calculations. More details about the reconstruction of states, in particular in the presence of noise, may be found in Refs. [18,19]. The reconstruction from interspike intervals is discussed by *Sauer* [20].

In the following we will assume that the scalar signal $s(t) = h(\mathbf{x}(t))$ is sampled with a sampling time t_s. The resulting time series $\{s^n\}$ with $s^n = s(nt_s)$ is used to reconstruct the states

$$\mathbf{y}^n = (s^n, s^{n+l}, s^{n+2l}, s^{n+(d-1)l}) \qquad (2)$$

for n = 1, ..., N. The symbol l denotes the delay time or lag in units of the sampling time (i.e., $t_l = lt_s$).

3.2 Estimating suitable reconstruction parameters

The reconstruction using delay coordinates is based on two parameters: the embedding dimension d and the delay time t_l. When the dimension d is chosen too small the conditions given in the embedding theorems are not fulfilled. For d too large, on the other hand, practical problems occur due to the fixed amount of points, that constitute thinner and thinner sets in $\mathbb{R}^d$ when d is increased. Concerning the choice of the delay time t_l the theorems exclude only a few discrete (non-generic) values. Thus, different t_l lead to reconstructions of the attractor that are diffeomorphically equivalent. They are, however, geometrically different. When the delay time t_l is too small, the coordinates $y_i^n = s^{n+(i-1)l}$ of each reconstructed state $\mathbf{y}^n$ do not significantly differ from one another and therefore the state points are scattered along the diagonal (Fig. 5a). In this case, any investigation of a possible fractal structure of the attractor (e.g., a dimension estimation) becomes difficult because one has to zoom into the cloud of points investigating scales that are (much) smaller than the diameter of the cloud perpendicular to the diagonal. On very small scales, however, such investigations become almost impossible due to the limited number of data points and the unavoidable presence of noise. The situation improves when t_l is increased. Then the attractor unfolds and its inner structure becomes "visible" already on larger scales as can be seen in Fig. 5b. When the delay time t_l is increased further, the reconstructed attractor is folded more and more (Fig. 5c). The origin of this folding is the "stretch and fold" mechanism that is typical for chaotic systems with sensitive dependence on initial conditions. Additional folds in the reconstruction are unwanted because they bring states close together in the reconstructed state space that are not close together in the original state space. In particular in the presence of noise, where state points are additionally shifted from the "right places", this may lead to cases where states are wrongly addressed. Even without noise, a too large value of t_l may lead to a folding that yields intersections of the attractor where the reconstruction is not one-to-one anymore. In this case, one may increase the dimension d to "repair" the embedding, but of course it is more sensible to decrease t_l. These examples show that a proper choice of t_l is crucial for any further investigation of the data.

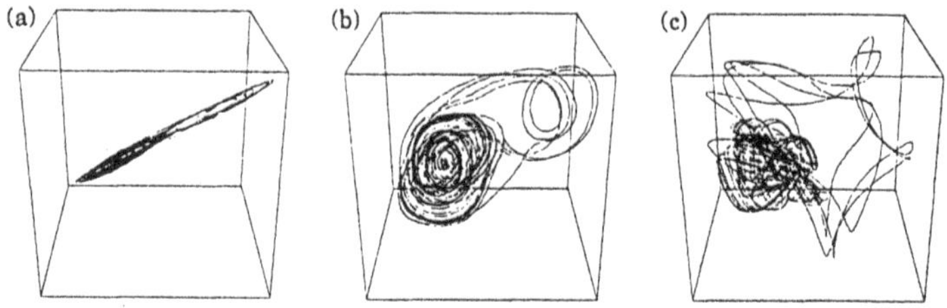

Figure 5: Three reconstructions with different delay times t_l for experimental data obtained from a chaotically oscillating, periodically driven pendulum.

Several algorithms have been suggested to find "optimal" values for d and t_l. Most of the methods for determining d are based on continuity tests for the induced flow in the reconstruction space [21-27] or for the embedding itself [28,29]. The main idea is to check whether closely neighbouring points are mapped to neighbouring points. The methods differ with respect to their concrete implementation. To find optimal values for t_l one class of algorithms directly considers the geometry of the reconstructed attractor, in particular its unfolding from the diagonal [30, 31]. The most popular approach consists in a minimization of the redundancy of the coordinates of the reconstructed states using linear autocorrelation functions or information theoretic concepts like mutual information [32-36]. Since bad values for t_l may destroy the embedding, the continuity tests for estimating d may also be used to determine t_l [22, 24-26, 28].

Until now, no method for determining optimal embedding parameters seems to be accepted as being the most useful. The reason for this diversity probably is the fact that there exists no pair (d, t_l) that is optimal for all possible subsequent investigations and applications. Furthermore, many of the algorithms proposed need as much computation time as a direct computation of a quantity of interest, e.g., the dimension. In this case, one may repeat the computation for several values of d and t_l and thus obtain not only "optimal" reconstruction parameters but also a test for the robustness of the method used and a first (quantitative) information about the data.

Finally, we want to mention the Broomhead-King-coordinates [37, 38] that result in a new coordinate system where the origin is shifted to the center of mass of the reconstructed states and the axes are given by the (dominant) principal components of the distribution of points. This new coordinate system is based on a Karhunen-Loève transformation that may be computed by a singular-value decomposition. A discussion of the advantages (e.g., noise reduction) and disadvantages of this "post-processing" of the reconstructed states may, for example, be found in Ref. [39].

3.3 Modelling and prediction

Once the states of the underlying dynamical system have been reconstructed it is possible to study their temporal evolution. Via the delay coordinates map F the flow ϕ^t in the original state space induces a flow $\psi^t = F \cdot \phi^t \cdot F^{-1}$ in the reconstructed state space. To model this induced flow different methods have been devised. An important class are local approximations where only states in a neighbourhood of a given reference point are used to compute the parameters of a locally valid model. Possible models are, for example, polynomials (including locally linear approximations) [40] or radial basis functions [41]. Of course, all methods for modelling the induced flow locally can in principle also be used as a global ansatz [41]. In general, local methods are more flexible and often yield better results when used for prediction of the time series. Global methods, on the other hand, provide model equations in "closed form" that may yield a better understanding of the underlying dynamics. When the time series originates from a continuous dynamical system and a global model is desired one should try to fit a vector field (differential equation) to the dynamics given by the reconstructed states. More details concerning different modelling techniques, typical problems and an interesting method for "verifying" a derived model using synchronization may be found in Ref. [42]. In any case, a successful approximation of the induced flow may not only be used for prediction but also provides a strong evidence for the deterministic nature of the data. Furthermore, good (local) models are a prerequisite for many other methods of nonlinear time series analysis and nonlinear dynamics.

3.4 Fractal dimensions

The (fractal) dimension of an attractor characterizes its complexity and gives a lower bound for the number of equations or variables needed for modelling the underlying dynamical process. From the infinite family of (generalized) dimensions D_q [43] the correlation dimension D_2 introduced by *Grassberger* and *Procaccia* [43,44] is mostly used in nonlinear time series analysis. This dimension is given by the scaling

$$C_d(r) \propto r^{D_2} \tag{3}$$

of the *correlation sum*

$$C_d(r) = \frac{2}{(N-c)(N-c+1)} \sum_{i=1}^{N} \sum_{j=1}^{i-c} H(r - \|y^i - y^j\|) \qquad (4)$$

that counts the relative number of neighbouring points closer than r (H is the Heaviside function with $H(x) = 1$ for $x > 1$ and zero elsewhere). The constant c is some correlation length and is used to omit points that are close neighbours in time [45], and d is the embedding dimension. The correlation sum approach is a *fixed size* method, because for a given radius r the number of pairs with distance smaller than r is counted. As an alternative one may also use a *fixed mass* method to estimate the dimension of the attractor as was pointed out by *Badii and Politi* [46, 47] and *Grassberger* [48]. In this case the m nearest neighbours of each reference point y^n and the radius $r^n = r(m)$ of this cloud of m points are determined. For the limit $m/N \to 0$ one obtains an approximation of the

$$\textit{capacity dimension}\ D_0 \approx -\frac{\log N}{\log \frac{1}{N} \sum_{n=1}^{N} r^n} \qquad (5)$$

and an estimate of the

$$\textit{information dimension}\ D_1 = -\frac{\log N}{\frac{1}{N} \sum_{n=1}^{N} \log r^n} \qquad (6)$$

To investigate the scaling in the limit $m/N \to 0$ one can decrease the number of neighbours m or increase the number of data points N.

Figure 6 gives an example of the fixed size method, where the correlation sum $C_3(r)$ of (4) is calculated and plotted versus the radius r of the ball of neighbouring points on a doubly logarithmic scale for the Duffing data of Fig. 3. The slope in a suitable intermediate range of r then gives the fractal dimension of the chaotic Duffing attractor (in the Poincaré section plane). The dashed line in Fig. 6 is a fit to the slope (shown with an offset) and gives the correlation dimension $D_2 = 1.1$ in the Poincaré plane, i. e., a fractal dimension of 2.1 for the continuous system. To detect whether the data are distinct from noise, surrogate data have been derived from the original data set according to the method of random Fourier phases as discussed below in section 3.7. Their correlation sum $C_3(r)$ is included in the figure as dotted curve. It is strongly distinct from the curve for the original data indicating a clear difference between the chaotic data and coloured noise. The error bars correspond to ±5 standard deviations.

A very suitable method is the determination of the *pointwise dimensions* and their average [49, 50]. A pointwise dimension D at a point of an attractor can be calculated by exploring the scaling behaviour of the "mass" of the attractor in the vicinity of the point, i. e., the number of points within a certain distance r. The number $N(r)$ of points in this ball of radius r should scale like $N(r) \propto r^D$ for $r \to 0$. When this is done for all points of the attractor, a histogram of pointwise dimensions is obtained. The average dimension and the standard deviation then are numbers to characterize the attractor. However, the histogram itself can be considered a characteristic of the attractor.

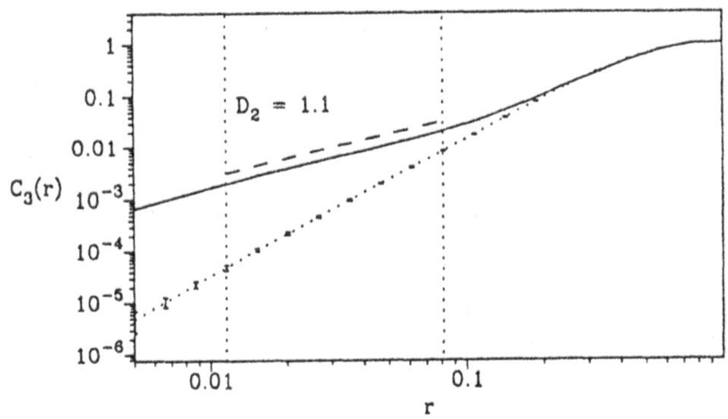

Figure 6: Correlation sum $C_3(r)$ vs. the distance r of neighbouring points for the Duffing data of Fig. 3. The dashed curve is a fit to the slope giving $D_2 = 1.1$. The dotted curve is from surrogate data to detect determinism.

When the attractor is sufficiently low-dimensional, its (fractal) dimension starting from low- dimensional embedding spaces ($d = 1, 2, 3,\ldots$) will level off at some value of d and does not increase further. This value then is said to determine the relevant number of (nonlinear) degrees of freedom, i. e., the number of dependent variables, of the dynamical system investigated. More details about dimension estimation methods, their possible pitfalls, extensions, and many references are given in the review articles [43, 51, 52].

3.5 Lyapunov spectra

The Lyapunov spectrum is a set of real numbers $\{\lambda_1, \lambda_2, \ldots, \lambda_m\}$, the Lyapunov exponents. Lyapunov exponents describe the mean exponential increase or decrease of small perturbations on an attractor and are invariant with respect to diffeomorphic changes of the coordinate system. When the largest Lyapunov exponent is positive, the system is said to be chaotic and to show sensitive dependence on initial conditions.

That way, the notion of chaos can be quantified. For an exact definition and computational details see, for example, Refs. [53-58].

The different methods for computing Lyapunov exponents from time series that have been proposed so far can be devided into two classes: *Jacobian-based methods* and *direct methods*. In the *Jacobian methods*, a model is first fitted to the data and the Jacobian matrices of the model equations are then used to compute the Lyapunov exponents using standard algorithms which have been developed for the case when the equations of the dynamical system are known [54-57]. Usually local linear approximations are used [59-67]. Modifications and improvements of the basic algorithm include:

- using neighbouring states from a shell ($r_{min} < r < r_{max}$) instead of a ball ($r < r_{max}$) to reduce the influence of noise (see, e. g., Refs. [60, 64, 65, 67])
- using nonlinear approximations of the flow by polynomials [68-71], radial basis functions [72-74] or neural networks [75]
- using projections onto linear subspaces based on local Broomhead-King coordinates [63]
- using the embedding dimension d for selecting the neighbours and a small *local dimension* d_L for performing the fit and computing the resulting d_L Lyapunov exponents (see, e.g., [60, 69, 70]).

The latter two methods have been devised to avoid so-called *spurious Lyapunov exponents,* that occur when the dimension d of the reconstructed state space is larger than the dimension m of the original state space. Spurious Lyapunov exponents do not depend on the dynamics but on the approximation scheme, the noise contained in the data and in some cases on the local curvature of the attractor. In principle they can take any values. To identify spurious Lyapunov exponents one can estimate the local thickness of the attractor along the directions associated with the different Lyapunov exponents [69, 70] or compare the exponents obtained with those computed for the time reversed series [66, 73], because spurious exponents correspond to directions where the attractor is very thin and because they do not change their signs upon time reversal (in contrast to the true exponents). An investigation of the data requirements for Jacobian-based methods may be found in Refs. [76, 77].

Direct methods directly estimate the divergent motion of the reconstructed states without fitting a model to the data. The best known and widely used method was introduced by *Wolf et al.* [58]. There, a reference orbit $\{\mathbf{y}^n\}$ is considered and a neighbouring orbit $\{\mathbf{z}^n\}$ starting from the nearest neighbour $\mathbf{z}^0$ of $\mathbf{y}^0$. After some transient the difference vector $\mathbf{u}^k = \mathbf{z}^{n+k} - \mathbf{y}^{n+k}$ points into the (time dependent) direction corresponding to the largest Lyapunov exponent λ_1. Then $\|\mathbf{u}^k\|$ grows, on the average, exponentially with $\exp(\lambda_1 t_s k)$ until it exceeds the range where a linear approximation of the flow at $\mathbf{y}^n$ holds. Now the neighbouring orbit $\{\mathbf{z}^n\}$ has to be replaced by another neighbouring orbit that is closer to the reference orbit *and* whose initial value lies on or near the line from the current reference state to the last point of the previous neighbouring orbit in order to preserve the direction corresponding to the largest Lyapunov exponent. Criteria for the replacement threshold and other details of the algorithm are given in Ref. [58], including a FORTRAN program. In principle, it is possible to use this strategy also for computing the second largest Lyapunov exponent [58], but this turnes out to be difficult. Thus it is quite fortunate that in many cases it suffices to establish the existence of at least one positive Lyapunov exponent. Due to its robustness the Wolf-algorithm is often used for the analysis of experimental data (see, for example, Refs. [78, 79]).

Another direct method, which is even simpler in terms of less free parameters, was proposed by *Sato et al.* [80] and *Kurths and Herzel* [81]. Very similar to the Wolf-algorithm, the average exponential growth of the distance of neighbouring orbits is studied on a logarithmic scale, this time via the prediction error

$$p(k) = \frac{1}{N t_s} \sum_{n=1}^{N} \log_2 \frac{\|\mathbf{y}^{n+k} - \mathbf{y}^{nn+k}\|}{\|\mathbf{y}^n - \mathbf{y}^{nn}\|} \tag{7}$$

where $\mathbf{y}^{nn}$ is the nearest neighbour of $\mathbf{y}^n$. The dependence of the prediction error $p(k)$ on the number of time steps k may be devided into three phases. Phase I is the transient where the neighbouring orbit converges to the direction corresponding to the largest Lyapunov exponent. During phase II the distance growths exponentially with $\exp(\lambda_1 t_s k)$ until it exceeds the range of validity of the linear approximation of the flow around the reference orbit $\{\mathbf{y}^{n+k}\}$. Then phase III begins where the distance increases slower than exponentially until it decreases again due to foldings in the state space. If phase II is sufficiently long, a linear segment with slope λ_1 appears in the $p(k)$ vs. k diagram. This not only allows an estimation of the largest Lyapunov exponent λ_1 but also provides a direct verification of exponential growth of distances to distinguish deterministic chaos from stochastic processes where a non-exponential separation of trajectories occurs [82]. Figure 7 gives an example for the determination of the largest Lyapunov exponent λ_1 by this method. The dotted curve is from surrogate data to detect determinism. The error bars are for ±5 standard deviations.

Several authors have investigated modifications of this basic algorithm [26, 83, 84]. In particular, those states must be excluded from the search for the nearest neighbour that are close together in time. Such states belong to the same segment of the trajectory and thus cannot converge to the most expanding direction. A possible criterion for a sufficiently large separation in time is the first minimum or zero of the autocorrelation function of the data set.

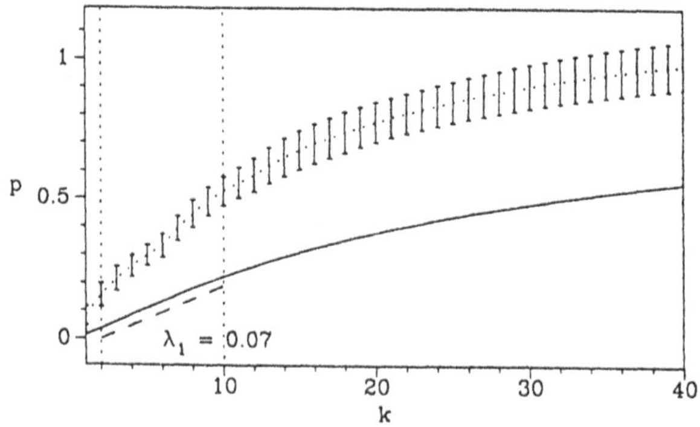

Figure 7: Prediction error p for experimental pendulum data in a Poincaré section plane vs. the delay number k. The slope of the solid line in the intermediate range of k gives the largest Lyapunov exponent λ_1.

3.6 Noise reduction

Experimental data usually are contaminated with noise. To improve noisy data two possibilities are available: Linear filters and special nonlinear noise reduction methods that make use of the deterministic origin of the data. The influence of linear filters on the reconstruction of states was investigated by *Sauer et al.* [16,17] and by *Broomhead et al.* [85]. Infinite impulse response (IIR) filters in general destroy the diffeomorphic equivalence between the original state space and the reconstructed state space. For finite impulse response (FIR) filters one can show that those filters which destroy the embedding constitute linear subspaces in the space of all FIR-filters. Therefore, with probability one, FIR-filters preserve the embedding. Nevertheless, one may find examples where the application of a FIR-filter makes subsequent estimations of Lyapunov exponents or dimensions more difficult as compared with the original "noisy" time series. Therefore, before filtering the data, one should always try to investigate the raw data. If it is known that the data originate from a deterministic system, methods for nonlinear noise reduction may be applied that have been developed only recently. The basic idea is to exploit the underlying determinism by correcting the states and the values of the time series using a (local or global) model that has previously been fitted to the (raw) data. More details about these applications of nonlinear time series analysis for noise reduction can be found in Refs. [86-88].

3.7 Detecting determinism

If the time series originates from a yet unknown process it is important to investigate the question whether the data contain some nonlinear deterministic dependencies or whether it is a purely stochastic signal, for instance, coloured noise. Several methods have been devised to answer this question [89-98]. Here, we want to focus on the *method of surrogate data* [89, 90] that provides a rather general and flexible framework for investigating unknown data. The basic idea is to make some hypothesis about the data and then to try to falsify this hypothesis. As an example we will use the widely used hypothesis: "The data are (just) coloured noise". In the first step, the data are modified in such a way that all structure except for the assumed properties will be destroyed or "randomized". This may be done, for example, by Fourier transforming the data and substituting the phases of the complex Fourier transform by random numbers. The power spectrum (or equivalently, the autocorrelation function) is not affected by this modification. After transforming back into the time domain we thus obtain a new time series with the same power spectrum. If the original data are just coloured noise we do not have destroyed any underlying structure, and subsequent estimations of dimensions, Lyapunov exponents, prediction errors, etc. should give the same results for the original time series and the surrogate data. If, however, the analysis yields significant differences, then our original time series was more than "just noise". To improve the statistical robustness of this test one usually generates several surrogate data sets and compares the mean values of the quantities obtained in the course of the subsequent data analysis with the corresponding value from the original time series. This has been done above in the examples, where the fractal dimension and the largest Lyapunov exponent were calculated for chaotic Duffing oscillator data and for experimental data from a driven pendulum, respectively. The surrogate data show strong deviations, indicating that the data investigated come from a deterministic, chaotic system. Furthermore, it is necessary to take into account a possibly performed static nonlinear transformation of the data that would distort the Gaussian distribution of the assumed coloured noise (Algorithm II in Ref. [89], see also Ref. [99]).

3.8 Synchronization of chaotic dynamics

Synchronization of periodic signals is a well-known phenomenon in physics, engineering and many other scientific disciplines [100-107]. However, even chaotic systems may be linked in a way such that their chaotic oscillations are synchronized. In particular, the case of onedirectional coupling has been investigated intensely during the last years [108-111] because of its potential application in communication systems [112-121]. There, an information signal containing a message is transmitted by using a chaotic signal as a broadband carrier, and the synchronization is necessary to recover the information at the receiver. Different implementations of this basic idea have been suggested, and we will discuss one of the most promising approaches [115-118] here.

The basic idea of this synchronization method consists in a decomposition of a given (chaotic) system into an active and a passive part, where different copies of the passive part synchronize when driven by the same active component. A general description of this *active-passive decomposition (APD)* and an example for illustration are given below.

Consider an arbitrary N-dimensional (chaotic) dynamical system

$$\dot{\mathbf{z}} = \mathbf{F}(\mathbf{z}). \tag{8}$$

The goal is to rewrite this autonomous system as a non-autonomous system that possesses certain synchronization properties. Formally, we may write

$$\dot{\mathbf{x}} = \mathbf{f}(\mathbf{x}, \mathbf{s}), \tag{9}$$

where $\mathbf{x}$ is the new state vector corresponding to $\mathbf{z}$ and s is some vector valued function of time given by

$$\mathbf{s} = \mathbf{h}(\mathbf{x}) \tag{10}$$

or

$$\dot{\mathbf{s}} = \mathbf{h}(\mathbf{x}, \mathbf{s}). \tag{11}$$

The pair of functions **f** and **h** constitutes a decomposition of the original vector field **F** (see also the example that follows). The crucial point of this decomposition is that for suitable choices of the function **h** any system

$$\dot{\mathbf{y}} = \mathbf{f}(\mathbf{y}, \mathbf{s}) \tag{12}$$

that is given by the *same* nonautonomous vector field **f**, the *same* driving s, but *different* variables **y**, synchronizes with the original system (9), i.e., $\|\mathbf{x} - \mathbf{y}\| \to 0$ for $t \to \infty$. More precisely, synchronization of the pair of (identical) systems (9) and (12) occurs if the dynamical system describing the evolution of the difference vector $\mathbf{e} = \mathbf{y} - \mathbf{x}$,

$$\dot{\mathbf{e}} = \mathbf{f}(\mathbf{y}, \mathbf{s}) - \mathbf{f}(\mathbf{x}, \mathbf{s}) = \mathbf{f}(\mathbf{x} + \mathbf{e}, \mathbf{s}) - \mathbf{f}(\mathbf{x}, \mathbf{s}),$$

possesses a stable fixed point at the origin $\mathbf{e} = 0$. In some cases, this can be proved via a stability analysis of the linearized system for small **e**,

$$\dot{\mathbf{e}} = D\mathbf{f}(\mathbf{x}, \mathbf{s}) \cdot \mathbf{e},$$

or by using (global) Lyapunov functions. In general, however, the stability has to be checked numerically using the fact that synchronization occurs if all conditional Lyapunov exponents [108] of the nonautonomous system (9) are negative.

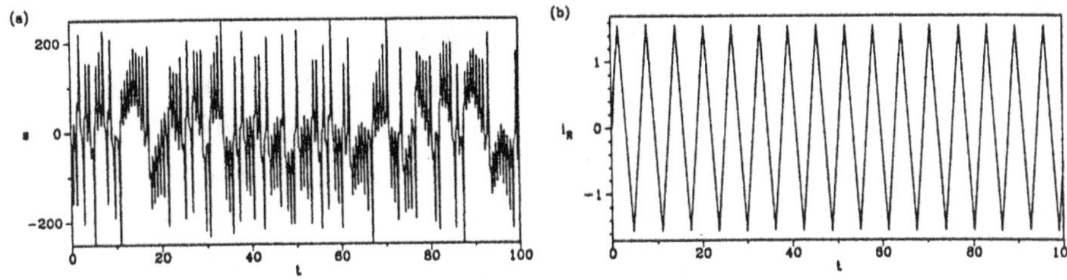

Figure 8: (a) Transmitted signal s consisting of an information signal modulated onto a chaotic carrier, (b) Recovered information signal i_R.

In this case, system (9) is a passive system and we call the decomposition an *active-passive decomposition (APD)* of the original dynamical system (8). The technical notion of conditional Lyapunov exponents was introduced by *Pecora and Carroll* [108] in order to study the synchronization of subsystems. The stability of the passive parts does not exclude chaotic solutions. To illustrate this feature we consider a decomposition of the Lorenz model [122].

$$\begin{aligned} \dot{x}_1 &= -10x_1 + s(t) & \dot{y}_1 &= -10y_1 + s(t) \\ \dot{x}_2 &= 28\,x_1 - x_2 - x_1x_3 & \dot{y}_2 &= 28\,y_1 - y_2 - y_1y_3 \\ \dot{x}_3 &= x_1x_2 - 2.666x_3 & \dot{y}_3 &= y_1y_2 - 2.666y_3 \end{aligned}$$

with

$$s(t) = 10x_2$$

To estimate the temporal evolution of the difference vector $\mathbf{e} = \mathbf{y} - \mathbf{x}$ of the states of the two systems we note that the difference $e_1 = y_1 - x_1$ of the first components converges to zero, because $e_1 = -10e_1$. Therefore, the remaining two-dimensional system describing the evolution of the differences $e_2 = y_2 - x_2$ and $e_3 = y_3 - x_3$ can, in the limit $t \rightarrow \infty$, be written as:

$$\dot{e}_2 = -e_2 - x_1(t)e_3$$

$$\dot{e}_3 = x_1(t)e_2 - 2.666e_3.$$

Using the Lyapunov function $L = e_2^2 + e_3^2$ it can be can shown that $\dot{L} = -2(e_2^2 + 2.666e_3^2) < 0$. This means, that the synchronization is globally stable and occurs for all types of driving signals $s(t)$.

What makes the method introduced above for constructing synchronized systems interesting for applications is the fact that in many cases the function $s(t)$ can be rather general. In particular, it may depend not only on the state $\mathbf{x}$ but also on some information signal $i(t)$, i.e., $s = h(\mathbf{x}, i)$. This feature can be used in a communication scheme, where $\dot{\mathbf{x}} = \mathbf{f}(\mathbf{x}, s)$ is the (chaotic) dynamical system of the transmitter, $s = h(\mathbf{x}, i)$ is the transmitted and received signal, and $\dot{\mathbf{y}} = \mathbf{f}(\mathbf{y}, s)$ constitutes the receiver. In the case of synchronization, i.e., when $\mathbf{x} = \mathbf{y}$, the information i can be recovered from $s = h(\mathbf{x}, i) = h(\mathbf{y}, i)$ if this equation is uniquely solvable for i.

As an example, we consider (13) with $s(t) = 10x_2 + i(t)x_3$. Since the variable x_3 is always positive, the information signal can be recovered exactly as $i_R(t) = (s - 10y_2)/x_3$ as soon as the transient of the synchronization process is over. Figure 8 shows the result of a numerical simulation where $i(t)$ is a triangular oscillation.

The communication method introduced above can also be implemented using high-dimensional systems and transmitted signals that are hyperchaotic, i. e., when more than one Lyapunov eponent are larger than zero. In this case, however, it is often difficult to find a region in parameter space where hyperchaos exists. An alternative are methods for synthesizing highdimensional systems in a systematic way, using standard low-dimensional systems with a known dynamics [115, 117].

4 Acknowledgments

The work has been sponsored by the Sonderforschungsbereich 185 - Nichtlineare Dynamik (nonlinear dynamics), the Fraunhofer-Gesellschaft, München, and further grants from the Deutsche Forschungsgemeinschaft. Thanks go to my numerous coworkers for extended times of discussion and exchange of ideas, in particular to the Nonlinear Dynamics Groups of the Drittes Physikalisches Institut, Göttingen, and of the Institut fur Angewandte Physik, Darmstadt.

References

[l] Ott, E. [1993], *Chaos in Dynamical Systems, Cambridge Cambridge*, University Press.

[2] Thompson, J. M. T., H. B. Stewart [1986], *Nonlinear Dynamics and Chaos*, Chichester, Wiley.

[3] Schuster, H. G. [1988], Deterministic Chaos, 2nd ed., Weinheim, VHC Publishers.

[4] Bergé, P., Y. Pomeau, C. Vidal [1984], *Order within Chaos: Towards a Deterministic Approach to Turbulence*, New York, John Wiley and Sons.

[5] Moon, F. C. [1992], *Chaotic and Fractal Dynamics*, New York, John Wiley and Sons.

[6] Gaponov-Grekhov, A. V., M. I. Rabinovich, [1992], *Nonlinearities in Action - Oscillations, Chaos, Order, Fractals*, Berlin, Springer.

[7] Lauterborn, W., J. Holzfuss, [1991], "*Acoustic chaos*", Int. J. Bifurcation and Chaos, 1, 13-26.

[8] Lauterborn, W., U. Parlitz, [1988], "*Methods of chaos physics and their application to acoustics*", J. Acoust. Soc. Am., 84,1975-1993.

[9] Abarbanel, H.D.I., R. Brown, J.J. Sidorowich, L.SH. Tsimring, [1993], "*The analysis of observed chaotic data in physical systems*", Rev. Mod. Phys., 65, 1331-1392.

[10] Guckenheimer, J., P. Holmes, [1983], *Nonlinear Oscillations, Dynamical Systems, and Bifurcations of Vector Fields*, New York, Springer.

[11] Feigenbaum, M. J., [1978], "Quantitative universality for a class of nonlinear transformations", *J. Stat. Phys.*, 19, 25-52.

[12] Lauterborn, W., T. Kurz, M. Wiesenfeldt, [1995] *Coherent Optics*, Berlin, Springer.

[13] Mettin, R., U. Parlitz, W. Lauterborn, [1993], "Bifurcation structure of the driven van der Pol oscillator", *Int. J. Bifurcation and Chaos*, 3, 1529-1555.

[14] Packard, N.H., J.P. Crutchfield, J.D. Farmer, R.S. Shaw [1980], "*Geometry from a time series*", Phys. Rev. Lett., 45, 712-716.

[15] Takens, F. [1981], "Detecting strange attractors in turbulence", in *Dynamical Systems and Turbulence*, eds. Rand, D.A. & Young, L.-S., Berlin, Springer, 366-381. 14

[16] Sauer, T., Y. Yorke, M. Casdagli [1991], "Embedology", *J. Stat. Phys.*, 65, 579-616.

[17] Sauer, T., J.A. Yorke [1993], "How many delay coordinates do you need ?", *Int. J. Bifurcation and Chaos*, 3, 737-744.

[18] Casdagli, M., S. Eubank, J.D. Farmer, J. Gibson [1991], "State space reconstruction in the presence of noise", *Physica D*, 51, 52-98.

[19] Gibson, J.F., J.D. Farmer, M. Casdagli, S. Eubank [1992], "An analytic approach to practical state space reconstruction" *Physica D*, 57,1-30.

[20] Sauer, T.[1994], "Reconstruction of dynamical systems from interspike intervals", *Phys. Rev.* Lett., 72, 3811-3814.

[21] Cenys, A., K. Pyragas [1988], "Estimation of the number of degrees of freedom from chaotic time series", *Phys. Lett.* A,129, 227-230.

[22] Buzug, Th., T. Reimers, G. Pfister [1990], "Optimal reconstruction of strange Attractors from purely geometrical arguments", *Europhys. Lett.*, 13, 605-610.

[23] Alecsic, Z. [1991], "Estimating the embedding dimension", *Physica D*, 52, 362-368.

[24] Buzug, Th., G. Pfister [1992], "Optimal delay time and embedding dimension for delaytime coordinates by analysis of the global static and local dynamical behavior of strange attractors", *Phys. Rev. A*, 45, 7073-7084.

[25] Gao, J., Z. Zheng [1993], "Local exponential divergence plot and optimal embedding of a chaotic time series", *Phys. Lett. A*, 181, 153-158.

[26] Gao, J., Z. Zheng [1994]. "Direct dynamical test for deterministic chaos and optimal embedding of a chaotic time series", *Phys. Rev. E*, 49, 3807-3814.

[27] Huerta, R., C. Santa Cruz, J.R. Dorronsore, V. Lopez [1995], "Local state-space reconstruction using averaged scalar products of dynamical-system flow vectors", *Europhys. Lett.*, 29, 13-18.

[28] Liebert, W., K. Pawelzik, H.G. Schuster [1991], "Optimal embeddings of chaotic attractors from topological considerations", *Europhys. Lett.*, 14, 521-526.

[29] Kennel, M.B., R. Brown, H.D.I. Abarbanel [1992], "Determining embedding dimension for phase-space reconstruction using a geometrical construction", *Phys. Rev. A*, 45, 3403-3411.

[30] Kember, G., A.C. Fowler [1993], "A correlation function for choosing time delays in phase portrait reconstructions", *Phys. Lett. A*, 179, 72-80.

[31] Rosenstein, M.T., J.J. Collins, C.J. De Luca [1994], "Reconstruction expansion as a geometry-based framework for choosing proper dealy times", *Physica D*, 73, 82-98.

[32] Frazer, A.M., H.L. Swinney [1986], "Independent coordinates in strange attractors from mutual information", *Phys. Rev. A*, 33, 1134-1140.

[33] Frazer, A.M. [1989], "Reconstructing attractors from scalar time series: a comparison of singular system and redundancy criteria", *Physica D*, 34, 391-404.

[34] Frazer, A.M. [1989], "Information and entropy in strange attractors", *IEEE Trans. Info. Theory*, 35, 245-262. 15

[35] Liebert, W., H.G. Schuster [1989], "Proper choice of the time delay for the analysis of chaotic time series", *Phys. Lett. A*, 142,107-111.

[36] Martinerie, J.M., A.M. Albano, A.I. Mees, P.E. Rapp [1992], "Mutual information, strange attractors, and the optimal estimation of dimension", *Phys. Rev. A*, 45, 7058-7064.

[37] Broomhead, D.S., G.P. King [1986], "Extracting qualitative dynamics from experimental data", *Physica D*, 20, 217-236.

[38] Landa, P.S., M.G. Rosenblum [1991], "Time series analysis for system identification and diagnostics", *Physica D*, 48, 232-254.

[39] Palus, M., I. Dvorak [1992], "Singular-value decomposition in attractor reconstruction: pitfalls and precautions", Physica D, 55, 221-234.

[40] Farmer, J.D., J.J. Sidorowich [1987], "Predicting chaotic time series", *Phys. Rev. Lett.*, 59, 845-848.

[41] Casdagli, M. [1989], "Nonlinear Prediction of chaotic time series", *Physica D*, 35, 335-356.

[42] Brown, R. N.F. Rulkov, E.R. Tracy [1994], "Modeling and synchronizing chaotic systems from time-series data", *Phys. Rev. E*, 49, 3784-3800.

[43] Grassberger, P., T. Schreiber, C. Schaffrath [1991], "Nonlinear time sequence analysis", *Int. J. Bifurcation and Chaos*, 1, 521-547.

[44] Grassberger, P., I. Procaccia [1983], "On the characterization of of strange attractors", *Phys. Rev. Lett.*, 50, 346-349.

[45] Theiler, J. [1986], "Spurious dimension from correlation algorithms applied to limited time-series data", *Phys. Rev. A*, 34, 2427-2431.

[46] Badii, R., A. Politi [1984], "Hausdorff dimension and uniformity factor of strange attractors", *Phys. Rev. Lett*, 52,1661-1664.

[47] Badii, R., A. Politi [1985], "Statistical description of chaotic attractors", *J. stat. Phys.*, 40, 725-750.

[48] Grassberger, P. [1985], "Generalizations of the Hausdorff dimension of fractal measures", *Phys. Lett. A*, 107, 101-105.

[49] Holzfuss, J., G. Mayer-Kress [1986], "An approach to error-estimation in the application of dimension algorithms", in [50], 114-122.

[50] Mayer-Kress, G. (ed.) [1986], Dimensions and Entropies in Chaotic Systems - Quantification of Complex Behavior, Berlin, Springer.

[51] Theiler, J. [1990], "Estimating fractal dimension", *J. Opt. Soc.* Am. A,7, 1055-1073.

[52] Broggi, G. [1988], "Evaluation of dimensions and entropies of chaotic systems", *J. Opt. Soc. Am. B*, 5, 1020-1028.

[53] Oseledec, V.I. [1968], "A multiplicative ergodic theorem. Lyapunov characteristic numbers for dynamical systems", *Trans. Moscow Math. Soc.*, 19, 197-231.

[54] Benettin, G., L. Galgani, A. Giorgilli, J.-M. Strelcyn [1980], "Lyapunov characteristic exponents for smooth dynamical systems and for hamiltonian systems; a method for computing all of them. Part II: Numerical application " *Meccanica*, 15, 21-30. 16

[55] Shimada, I., T. Nagashima [1979], "A numerical approach to ergodic problems of dissipative dynamical systems", *Prog. Theor. Phys.*, 61, 1605-1616.

[56] Eckmann, J.-P., D. Ruelle [1985], "Ergodic theory of chaos and strange attractors", *Rev. Mod. Phys.*, 57, 617-656.

[57] Geist, K., U. Parlitz, W. Lauterborn [1990], "Comparison of Different Methods for Computing Lyapunov Exponents", *Prog. Theor. Phys.*, 83, 875-893.

[58] Wolf, A., J.B. Swift, L. Swinney, J.A. Vastano [1985], "Determining Lyapunov exponents from a time series", *Physica D*, 16, 285-317.

[59] Sano, M.,Y. Sawada [1985], "Measurement of the Lyapunov spectrum from a chaotic time series", *Phys. Rev. Lett.*, 55,1082-1085.

[60] Eckmann, J.-P., S.O. Kamphorst, D. Ruelle, S. Ciliberto [1986], "Lyapunov exponents from time series", *Phys. Rev. A*, 34, 4971-4979.

[61] Stoop, R.,P.F. Meier [1988], "Evaluation of Lyapunov exponents and scaling functions from time series", *J. Opt. Soc. Am. B*, 5, 1037-1045.

[62] Holzfuss, J.,W. Lauterborn [1989], "Liapunov exponents from a time series of acoustic chaos", *Phys. Rev. A*, 39, 2146-2152.

[63] Stoop, R.,J. Parisi [1991], "Calculation of Lyapunov exponents avoiding spurious elements", *Physica D*, 50, 89-94.

[64] Zeng, X., R. Eykholt, R.A. Pielke [1991], "Estimating the Lyapunov-exponent spectrum from short time series of low precision", *Phys. Rev. Lett.*, 66, 3229-3232.

[65] Zeng, X., R.A. Pielke, R. Eykholt [1992], "Extracting Lyapunov exponents from short time series of low precision", *Modern Phys. Lett. B*, 6, 55-75.

[66] Parlitz, U. [1993], "Lyapunov exponents from Chua's circuit", J. Circuits, Systems and Computers, 3, 507-523.

[67] Kruel, Th.M., M. Eiswirth, F.W. Schneider [1993], "Computation of Lyapunov spectra: Effect of interactive noise and application to a chemical oscillator", *Physica D*, 63, 117-137.

[68] Briggs, K. [1990], "An improved method for estimating Liapunov exponents of chaotic time series", *Phys. Lett. A*, 151, 27-32.

[69] Bryant, P., R. Brown, H.D.I. Abarbanel [1990], "Lyapunov exponents from observed time series", *Phys. Rev. Lett.*, 65,1523-1526.

[70] Brown, R., P. Bryant, H.D.I. Abarbanel [1991], "Computing the Lyapunov spectrum of a dynamical system from an observed time series", *Phys. Rev. A*, 43, 2787-2806.

[71] Abarbanel, H.D.I., R. Brown, M.B. Kennel [1991], "Lyapunov exponents in chaotic systems: their importance and their evaluation using observed data", *Int. J. Mod. Phys. B*, 5, 1347-1375.

[72] Holzfuss, J, U. Parlitz [1991], "Lyapunov exponents from time series", Proceedings of the Conference Lyapunov Exponents, Oberwolfach 1990, eds. L. Arnold, H. Crauel, J.-P. Eckmann, in: *Lecture Notes in Mathematics*, Springer Verlag.

[73] Parlitz, U. [1992], "Identification of true and spurios Lyapunov exponents from time series", *Int. J. Bifurcation and Chaos, 2*, 155-165. 17

[74] Kadtke, J.B., J. Brush, J. Holzfuss [1993], "Global dynamical equations and Lyapunov exponents from noisy chaotic time series", *Int. J. Bifurcation Chaos*, 3, 607-616.

[75] Gencay, R., W.D. Dechert [1992], "An algorithm for the n Lyapunov exponents of an n-dimensional unknown dynamical system", *Physica D*, 59, 142-157.

[76] Eckmann, J.-P., D. Ruelle [1992], "Fundamental limitations for estimating dimensions and Lyapunov exponents in dynamical systems", *Physica D*, 56, 185-187.

[77] Ellner, S., A.R. Gallant, D. McCaffrey, D. Nychka [1991], "Convergence rates and data requirements for Jacobian-based estimates of Lyapunov exponents from data", *Phys. Lett. A*, 153,357-363.

[78] Fell, J., J. Roschke, P. Beckmann [1993], "Deterministic chaos and the first positive Lyapunov exponent: a nonlinear analysis of the human electroencephalogram during sleep", *Biol. Cybern.*, 69, 139-146.

[79] Fell, J., P. Beckmann [1994], "Resonance-like phenomena in Lyapunov calculations from data reconstructed by the time-delay method" *Phys. Lett. A*, 190, 172-176.

[80] Sato, S., M. Sano, Y. Sawada [1987], "Practical methods of measuring the generalized dimension and largest Lyapunov exponent in high dimensional chaotic systems", *Prog. Theor. Phys.*, 77, 1-5.

[81] Kurths, J., H. Herzel [1987], "An attractor in solar time series", *Physica D*, 25,165-172.

[82] Dammig, M., F. Mitschke [1993], "Estimation of Lyapunov exponents from time series: the stochastic case", *Phys. Lett. A*, 178, 385-394.

[83] Rosenstein, M.T., J.J. Collins, C.J. de Luca [1993], "A practical method for calculating largest Lyapunov exponents from small data sets", *Physica D*, 65,117.

[84] Kantz, H. [1994], "A robust method to estimate the maximal Lyapunov exponent of a time series", *Phys. Lett. A*, 185, 77-87.

[85] Broomhead, D.S., J.P. Huke, M.R. Muldoon [1992], "Linear filters and nonlinear systems", *J. Roy. Stat. Soc.*, B54, 373-382.

[86] Grassberger, P., R. Hegger, H. Kantz, C. Schaffrath, T. Schreiber [1993], "On noise reduction methods for chaotic data" *CHAOS*, 3,127-141.

[87] Kantz, H., T. Schreiber, I. Hoffmann, T. Buzug, G. Pfister, C.G. Flepp, J. Simonet, R. Badii, E. Brun [1993], "Nonlinear noise reduction: A case study on experimental data", *Phys. Rev. E*, 48,1529-1538.

[88] Kostelich, E.J.,T. Schreiber [1993], "Noise reduction in chaotic time-series data: A survey of common methods", *Phys. Rev. E*, 48, 1752-1763.

[89] Theiler, J., B. Galdrikian, A. Longtin, S. Eubank, J.D. Farmer [1992], "Using surrogate data to detect nonlinearity in time series" in: Nonlinear Modeling and Forecasting, eds. M. Casdagli and S. Eubank, SFI Studies in the Sciences of Complexity, Vol.XII (Reading, MA,Addison-Wesley) 163-188.

[90] Theiler, J., S. Eubank, A. Longtin, B. Galdrikian, J.D. Farmer [1992], "Testing for nonlinearity in time series: the method of surrogate data", *Physica D*, 58, 77-94. 18

[91] Provenzale, A., L.A. Smith, R. Vio, G. Murante [1992], "Distiguishing between lowdimensional dynamics and randomness in measured time series", *Physica D*, 58, 31-49.

[92] Smith, L. [1992], "Identification and prediction of low dimensional dynamics", *Physica D*, 58, 50-76.

[93] Takens, F. [1993], "Detecting nonlinearities in stationary time series", *Int. J. of Bifurcation and Chaos*, 3, 241-256.

[94] Wayland, R., D. Bromley, D. Pickett, A. Passamante [1993], "Recognizing determinism in a time series", *Phys. Rev. Lett.*, 70, 580-582.

[95] Palus, M., V. Albrecht,I. Dvorak [1993], "Information theoretic test for nonlinearity in time series", *Phys. lett. A*, 175, 203-209.

[96] Kaplan, D. [1994], "Exceptional events as evidence for determinism", *Physica D*, 73, 38-48.

[97] Salvino, L.W., R. Cawley [1994], "Smoothness implies determinism: a method to detect it in time series", *Phys. Rev. Lett.*, 73, 1091-1094.

[98] Savit, R.M. Green [1991], "Time series and dependent variables", *Physica D*, 50, 95-116.

[99] Rapp, P.E., A.M. Albano, I.D. Zimmerman, M.A. Jiménez-Moltano [1994], "Phaserandomized surrogates can produce spurious identifications of non-random structure", *Phys. Lett. A*, 192, 27-33.

[100] Fujisaka, H.,T. Yamada [1993], "Stability theory of synchronized motion in coupled oscillator systems," *Prog. Theor. Phys.*, 69, 32-46.

[101] Singer, W. [1993], "Synchronization of cortical activity and its putative role in information processing and learning," *Annu. Rev. Physiol.*, 55, 349-374.

[102] Ashwin, P., J. Buescu, I. Stewart [1994], "Bubbling of attractors and synchronisation of chaotic oscillators," *Phys. lett. A*, 193,126-139.

[103] Heagy, J.F., T.L. Carroll, L.M. Pecora [1994], "Synchronous chaos in coupled oscillator systems," *Phys. Rev. E*, 50, 1874-1885.

[104] Lai, Y.-C., C. Grebogi [1994], "Synchronization of spatiotemporal chaotic systems by feedback control," *Phys. Rev. E*, 50,1894-1899.

[105] Collins, J.J.,I. Stewart [1994], "A group-theoretic approach to rings of coupled biological oscillators," *Biol. Cybern.*, 71, 95-103.

[106] Lindner, J.F., B.K. Meadows, W.L. Ditto, M.E. Inchiosa, A.R. Bulsara [1995], "Array enhanced stochastic resonance and spatiotemporal synchronization," *Phys. Rev. Lett.*, 75, 3-6.

[107] Braiman, Y., W.L. Ditto, K. Wiesenfeld, M.L. Spano [1995], "Disorder-enhanced synchronization," Phys. Lett. A 206, 54-60; Braiman, Y., J.F. Lindner, W.L. Ditto [1995], "Taming spatiotemporal chaos with disorder," *Nature*, 378, 465-467.

[108] Pecora, L.M., T.L. Carroll [1990], "Synchronization in chaotic systems," *Phys. Rev. Lett.*, 64, 821-824; Carroll, T.L., L.M. Pecora [1991], *IEEE Trans.* Circuits Syst. 38, 453-456; Carroll T.L., L.M.Pecora [1992], *Int. J. Bif. Chaos*, 2, 659-667; Carroll, T.L., L.M.Pecora [1993], *Physica D 67*, 126-140; Carroll, T.L. [1994], *Phys. Rev. E*, 50, 25802587. 19

[109] Rulkov, N.F., K.M. Sushchik, L.S. Tsimring, H.D.I. Abarbanel, [1995], "Generalized synchronization of chaos in directionally coupled chaotic systems," *Phys. Rev. E*, 51, 980-995.

[110] Brown, R., N.F. Rulkov, E.R. Tracy [1994], «Modelling and synchronizing chaotic systems from experimental data," *Phys. Lett. A*, 194, 71-76.

[111] U. Parlitz, U. [1996], "Estimating model parameters from time series by autosynchronization," *Phys. Rev. Lett.*, 76, 1232-1235

[112] Cuomo, K.M., A.V. Oppenheim [1993], "Circuit implementation of synchronized chaos with applications to communications," *Phys. Rev. Lett.*, 71, 65-68; Oppenheim, A.V. G.W.Wornell, S.H.Isabelle, K.M. Cuomo [1992], "Signal processing in the context of chaotic signals," *Proc IEEE ICASSP*, IV 117-120.

[113] Kocarev, L., K.S. Halle, K. Eckert, L.O. Chua, U. Parlitz [1992], "Experimental demonstration of secure communications via chaotic synchronization," *Int. J. Bifurcation and Chaos*, 2, 709-713.

[114] Parlitz, U., L.O. Chua, L. Kocarev, K.S. Halle, A. Shang [1992], "Transmission of digital signals by chaotic synchronization," *Int. J. Bifurcation and Chaos*, 2, 973-977.

[115] Wu, C.W., L.O. Chua [1993], "A simple way to synchronize chaotic systems with applications to secure communication systems," *Int. J. of Bifurcation and Chaos*, 3, 1619-1627; Wu, C.W., L.O. Chua [1994], "A unified framework for synchronization and control of dynamical systems," *Int. J. Bifurcation and Chaos*, 4, 979-998.

[116] Halle, K.S., C.W. Wu, M. Itoh, L.O. Chua [1993], "Spread spectrum communication through modulation of chaos," *Int. J. Bifurcation and Chaos*, 3, 469-477.

[117] Kocarev, L., U. Parlitz [1995], "General approach for chaotic synchronization with applications to communcation," *Phys. Rev. Lett.*, 74, 5028-5031; "Communication using chaotic signals," in: Proceedings of Nonlinear Dynamics of Electronic Systems, Krakow, Poland, 29 - 30 July, 1994.

[118] Parlitz, U., L. Kocarev, T. Stojanovski, H. Preckel [1996], "Encoding messages using chaotic synchronization," to appear in: *Phys. Rev. E.*

[119] Parlitz, U. L. Kocarev [1996], "Multichannel communication using auto-synchronization," to appear in: *Int. J. Bifurcation and Chaos.*

[120] Short, K.M. [1994], "Steps toward unmasking secure communications," *Int. J. Bifurcation and Chaos*, 4, 959-977.

[121] Pérez, G., H.A. Cerdeira [1995], "Extracting messages masked by chaos," *Phys. Rev. Lett.*, 74, 1970-1973.

[122] Lorenz, E.N. [1963], "Deterministic nonperiodic flow", *J. Atmos. Sci.*, 20,130-141.

LIGHTWAVE COMMUNICATIONS

Michel Joindot

1 Introduction

The history of modern lightwave communications systems began at the end of the fifties, with lasers, which were the first optical source with high power and directivity able to be used in a communication system. Gas lasers, needing high supply voltages and relatively large sized, were followed in 1961 by semiconductor lasers, which appeared much more promising, because of their small dimensions and the possibility of modulating them with low currents. More than ten years of research were necessary to achieve reliable enough semiconductor lasers working at ambient temperature.

Free space optical transmission did not appear as promising because, contrary to what happens in radio, attenuation due to hydrometeors leads to outage times uncompatible with the requirements of a telecommunication system.

The first research about optical fibers began around 1966, and at the beginning of the seventies, many laboratories were very active in this field. Constant progress were realized, allowing to achieve attenuations of 20 dB/km in 1975 and 0,2 dB/km in 1984. This last values opened the way to the implementation of practical systems competing with the other existing, especially those using coaxial cables, in terms of repeater spacing as well as capacity. The first fibers to be used were multimode fibers, having a relatively large diameter and supporting then the propagation of many electromagnetic modes. Later single mode fibers, i.e dielectric waveguides supporting only one mode, replaced multimode fibers, allowing very high bandwidths. Three so-called "transmission windows" are used in optical telecommunications. The first one, around 0.8 μm, was historically the first to be used, because the first semiconductor lasers operated at this wavelength. But it is no more employed, at least for high capacity transmission systems. Two other windows are still currently in use now, the first around 1.3 μm, where the material chromatic dispersion is minimum, the second around 1.55 μm, where the attenuation is minimum.

The first generation of lightwave systems, using the 0.8 μm window, began to be operated at the end of the seventies : they used multimode fibers and could transmit typically 50 Mbit/s, with a regenerating repeater spacing of roughly 10 km. Between 1980 and 1990, lightwave communication systems at 1.3 and 1.55 μm over singlemode fibers carrying hundreds of Mbit/s were introduced into the trunk network, with a repeater spacing up to 50 km, compared to 2 km for coaxial cable systems with the same bitrate. It became evident that optical fibers could compete successfully not only with metallic cables, but also with radio relay systems, which could not provide such a high capacity. At the beginning of the nineties, doped fiber amplifiers replacing the electronic regenerators, providing the possibility of in line amplification of the optical signal, became within a relatively short time industrial and reliable devices and then opened the way to very long haul systems, without electronics, and transparent to the bitrate, with some counterparts, for example noise accumulation along the line. They found straightforward applications in undersea lightwave systems.

Figure 1 shows the attenuation of a single mode fiber in dB/km versus the wavelength.

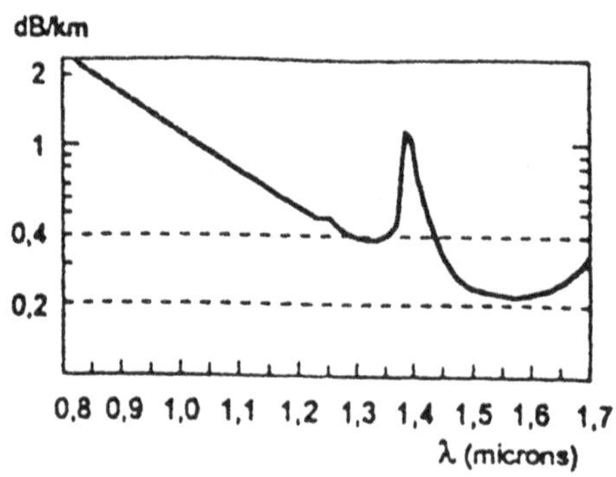

Figure 1 : Attenuation of a singlemode fiber versus wavelength.

2 The transmission medium

The single mode fiber itself was sometimes considered as a perfect low attenuation transmission medium, with such a large bandwidth that signals could be transmitted through it without suffering any distortion : this was true as long as the bit rates were relatively low and attenuation was the limiting factor of the transmission distance. In fact, signals propagating through the fiber are subject to distortions, linear distortions due to the transfer function of the fiber as well as non linear distortions caused in particular by the dependence between the refractive index and the power launched into the fiber through Kerr effect.

2.1 Distortion induced by chromatic dispersion

Linear distorsions are due to chromatic dispersion [*Agrawal*, 1989, *Vassallo*, 1991]. Inside the bandwidth occupied by the transmitted signals around an angular frequency ω_s, phase can be expanded to the second order according to the formula :

$$\Phi(\omega)=\Phi(\omega_s)+\tau(\omega_s)L.(\omega-\omega_s)+\frac{\beta_2}{2}L.(\omega-\omega_s)^2 \tag{1}$$

$\tau(\omega_s)$ is the group delay per unit length at angular frequency ω_s and β_2 the second derivative of the phase per unit length characterizes group delay variations versus ω. Instead of β_2, one uses the chromatic dispersion parameter $D(\lambda_s)$, expressed in ps/(nm.km), defined as the group delay variation for a length of 1 km and within a bandwidth corresponding to a wavelength variation of 1 nm around the wavelength λ_s.

$$D(\lambda_s) = -(6\pi \times 10^5)\frac{\beta_2(\lambda_s)}{\lambda_s^2} \tag{2}$$

where $\beta_2(\lambda_s)$ and λ_s are respectively expressed in ps²/km and nm. For the so called standard single mode fibers (SSF) at 1.55 µm, $\beta_2(\lambda_s)$ and $D(\lambda_s)$ equal respectively -20 ps²/km and 17 ps/(nm.km). An important parameter of the fiber is the zero dispersion wavelength λ_0, where $\beta_2(\lambda_0)$ cancels. For standard singlemode fibers, λ_0 is around 1.3 µm and dispersion is then much smaller around 1.3 µm, ensuring that propagating signals suffer less distortion, which explains why the 1.3 µm window has been used. But the counterpart is a larger attenuation than in the 1.55 µm window. Dispersion Shifted Fibers (DSF) exhibit around 1.55 µm a dispersion much smaller than standard fibers, typically less than 3.5 ps/(nm.km) in absolute value.

An impulse propagating along a fiber will then be distorted and broadened, which results into mutual overlapping when successive impulses are transmitted, and then into intersymbol interference, a universal impairement encountered in all the digital communication systems. Chromatic dispersion will then limit the bitrate for a given distance, or the distance for a given bitrate. Computer calculations are needed to get distortion suffered by the signal. Nevertheless, in the particular case of a gaussian impulse, computation can be carried out analytically. The envelope widens and remains gaussian, while the signal suffers also a phase modulation. The complex envelopes of the input and output signal after propagation over a length L of fiber are respectively expressed as [*Agrawal*, 1989] :

$$u(t,0) = A\exp\left(-\frac{t^2}{2\theta_0^2}\right)$$

$$u(t,L) = \frac{A}{\sqrt[4]{1+x^2}}\exp\left(-\frac{t^2 - i\,sgn(\beta_2).x\,t^2}{2\theta_0^2(1+x^2)} - \frac{i}{2}\arctan x\right),\ x = \frac{L}{L_D}\quad L_D = \frac{\theta_0^2}{|\beta_2|}, \tag{3}$$

The so-called dispersion length L_D, defined as the length after which the initial width is multiplied by $\sqrt{2}$ and depending on the initial width, gives an idea of the distance limitation induced by chromatic dispersion for a given bitrate. For instance, assuming standard fiber (D=17 ps/(nm.km)) and θ_0 equal to the half of the symbol time, L_D (in kilometers) is given by 12 500/B^2 with the bit rate B expressed in Gbit/s. Dispersion length values are then 500 and 125 km for B equal to 2.5 and 10 Gbit/s respectively.

It could be concluded that dispersion must be as low as possible in order to have the largest transmission distance : in fact, we will see in the next section that non linear effects become more and more important when the absolute value of chromatic dispersion decreases. Moreover, when β_2 is close to zero within the bandwidth of the transmitted signal, third order terms in the expansion of $\Phi(\omega)$ must be taken into account.

2.2 Distortion induced by non linear effects

In radio systems (for example satellite communications) non linear distortions suffered by signals are essentially localized in the power amplifiers working in non linear regime and the propagation medium is in most cases perfectly linear ; on the contrary non linearities in optical communication systems occur in the fiber itself when the launched power is sufficiently high. Four effects are due to the dependence of refractive index versus power (Kerr effect). Stimulated Brillouin and Raman scattering (SBS and SRS) which will be briefly described result from the interaction with phonons.

2.2.1 Self phase modulation (SPM)

If $u(t,0)$ is the complex envelope of the signal at the input of an ideal dispersionless fiber of length L, the complex envelope of the output signal can be written as [*Agrawal*, 1989] :

$$u(t,L) = u(t,0)\exp\left(-\frac{\alpha L}{2} + i\,\Phi(t,L)\right), \tag{4}$$

$$\text{with}\quad \Phi(t,L) = \gamma|u(t,0)|^2\,\frac{1-e^{-\alpha L}}{\alpha} = \gamma|u(t,0)|^2\,L_{eff},$$

where α is the attenuation per unit length, L_{eff} is the effective length (equal to the actual length in a lossless fiber) and γ a coefficient characterizing the non linearity, equal around to 3 W^{-1}/km. It is to be noted that γ depends of the effective mode area, i.e the area over which the energy is concentrated, which is typically between 50 and 80 μm^2 at 1.55 µm : the energy surface density is then a crucial parameter.

The maximum phase shift Φ_{max} occurs at the impulse center (t=0) and is given by :

$$\Phi_{max}(L) = |u(0,0)|^2\,\gamma\frac{1-e^{-\alpha L}}{\alpha} = \gamma\,P_0\frac{1-e^{-\alpha L}}{\alpha} = \frac{L_{eff}}{L_{NL}} \quad\text{with}\quad L_{NL} = \frac{1}{\gamma|u(0,0)|^2} = \frac{1}{\gamma P_0}, \tag{5}$$

where P_0 is the peak power and L_{NL} the characteristic length, corresponding to a phase shift of one radian.

Comparing the length L with L_D and L_{NL} shows whether the system performance is limited by dispersion or non linear effects. For instance, in the case of terrestrial communication systems with a typical launched power of 0 dBm (1 mW) L_{NL} equals 330 km, and assuming an attenuation of 0.2 dB/km (α = 0.046 km^{-1}) the asymptotic phase shift after 330 km is only 0.06 radian. These orders of magnitude prove that self phase modulation can be neglected in the case of these systems. On the contrary, non linear distortions play an essential role in the case of undersea lightwave systems, due to the very important length (several thousands of kilometers), or even on shorter distances when a booster amplifier, allowing to launch 15 dBm or more into the fiber is employed.

Combined effects of chromatic dispersion and self phase modulation will modify the impulse shape, and the resulting effect will depend strongly on the sign of β_2.

2.2.2 Cross Phase Modulation (XPM)

When several optical carriers propagate simultaneously in a fiber, each of them undergoes not only the previously described SPM effect, but also another phase modulation, called cross phase modulation (XPM), due to the fact that index perturbation is induced by the total power propagating in the fiber. Description of this phenomenon is very complex : analytical formulas can be found assuming pure unmodulated carriers or a dispersionless fiber, and show that XPM depends on the total power of all the carriers, which means that this effect is dominant compared to SPM. But, in the actually interesting case of modulated carriers and chromatic dispersion, results can only be obtained using complex computer softwares. It is interesting to notice that phase distortion due to XPM is roughly proportional to twice the total power, which means that its efficiency, expressed in degrees per mW, is twice larger than SPM efficiency.

2.2.3 Four Wave Mixing (FWM)

This effect also occurs when several carriers propagate simultaneously in a fiber. It generates intermodulation products and is through its effects very similar to what happens in multichannel radio systems. For instance, third order non linearity creates beat signals between three carriers at angular frequencies ω_1, $\omega_2 = \omega_1 + \Delta\omega$ and $\omega_3 = \omega_1 + 2\Delta\omega$ and creates then intermodulation products at angular frequencies $p\omega_1 + q\omega_2 + r\omega_3$ where p, q, r are integers such that $|p| + |q| + |r|$ equals 3. Obviously, the intermodulation product at angular frequency $2\omega_2 - \omega_1 = \omega_3$ will perturb carrier at angular frequency ω_3. Analytical formulas exist in the case of pure carriers [*Tkach*, 1995]. But, as for XPM, phenomena are much more complex in the practical case of modulated channels. FWM influence depends on channel spacing and fiber dispersion: highly dispersive fiber will reduce intermodulation through phase mismatch of neighbouring channels since the corresponding signals propagate with different group velocity. This is why low dispersion fibers are less favourable as far as FWM is concerned, especially for channels near the zero dispersion wavelength.

2.2.4 Modulation instability

If an optical carrier with power P_0 at angular frequency ω_s launched into a fiber is perturbed through another signal at frequency $\omega_s + \Omega$, non linear interaction can amplify the perturbing signal at the expense of carrier. This phenomenon exists only in the anomalous propagation regime (β_2 negative) and if Ω falls within a bandwidth defined by the condition [*Agrawal*, 1989] :

$$\Omega^2 \leq \Omega_c^2 = \frac{4\gamma P_0}{|\beta_2|} \tag{6}$$

where γ has been defined in (4). The maximum gain factor occurs for $\Omega = \pm\, \Omega_c/\sqrt{2}$ and equals $2\gamma P_0$ (in m^{-1}). Practically, modulation instability can amplify for instance noise around the carrier, causing carrier to noise degradation. When the carrier is intensity modulated, modulation instability splits the impulses into several narrower impulses and may completely destroy the information carrying signal. This is why positive dispersion is always avoided in undersea lightwave systems operating at very low dispersion (typically 0.1 ps/(nm.km)).

2.2.5 *Stimulated Brillouin and Raman scattering*

The basic physical reason is the energy transfer from the optical field to the non linear medium, i.e the fiber, exciting vibrational modes of silica, the basic difference between the two effects being that acoustic (resp. optical) phonons are involved in SBS (resp. SRS). They are governed by the set of differential equations representative of a parametric amplification process [*Agrawal*, 1989] :

$$\frac{dP_b}{dz} = gP_b\, P_p - \alpha P_b \qquad \frac{dP_p}{dz} = -gP_b P_p - \alpha P_p, \tag{7}$$

where P_b and P_p are respectively the probe and pump power, α the transmission factor of the fiber and g a gain characteristic of the phenomenon under consideration. The useful signal, acting as the pump, transfers its power to an interferer, denoted in (7) as the probe. We will just very briefly review the degradations due to both effects in optical communication systems, beginning by SBS.

When the pump power exceeds the Brillouin threshold (typically 1 to 3 mW in usual fibers), a backward propagating wave (probe), down shifted in frequency by 11 GHz in silica fibers and called Stokes wave, is generated at the expense of the signal acting as a pump. This will cause an additional attenuation of the useful signal and also harmful effects due to this counterpropagating wave coming back to the emitter. Maximum value of gain g_B is around 5×10^{-11} m^{-1}/W. The aforementioned threshold power was obtained for an unmodulated pump and increases when the pump bandwidth $\Delta\nu_p$ increases beyond the Brillouin gain bandwidth, typically 100 MHz, which means that frequency spreading due to modulation is a favourable factor. If a very high power is launched into the fiber, SBS can be prevented by a very low frequency modulation (dithering) of the laser, which broadens the linewidth without affecting the information signal.

In SRS, the frequency shift between pump and (here copropagating) probe is much larger (13 THz), amplification bandwidth is around 8 THz, but the maximum gain g_R is much smaller than for SBS, typically 10^{-13} m^{-1}/W. The SRS power threshold is then much higher, around 300 to 600 mW for silica fiber at 1.55 µm. SRS is then not to consider for most of the optical communication systems.

2.3 Polarization Mode Dispersion

In a perfectly circular and isotropic optical fiber, each electromagnetic mode is degenerated because of the existence of two independent states of polarization. In a polarization maintaining fiber, for instance an elliptical section fiber, geometrical axes define two linearly polarized modes propagating at different velocity and if a light impulse is injected into the fiber, the field components onto these two states propagate at different speed, leading to two impulses at the receiving end, with an arrival time difference proportional to fiber length and velocity difference.

An actual fiber is neither perfect, nor polarization maintaining : ellipticity or external constraints induce birefringence, with axes changing randomly along the fiber. Random mode coupling occurs along the fiber and it can be shown that two impulses are recovered at the detector output, separated by a random delay. This delay $\Delta\tau$ follows a Maxwell probability density function (this is a random tridimensional problem) in a perfectly random fiber, i.e when the length is large compared to the correlation length, defined as the length over which fiber characteristics do not change significantly [*Foschini*, 1991].

$\Delta\tau$ is usually characterized by its average value $<\Delta\tau>$, which is proportional to the square root of the fiber length : this parameter usually called PMD, is expressed in $ps/\sqrt{km}$. In digital transmission systems, polarization mode dispersion create intersymbol interference [*Poole*, 1991] and then bit error rate (BER) degradation increasing with $\Delta\tau$. The outage time, i.e the proportion of time during which BER exceeds the maximum tolerable value, increases with $<\Delta\tau>/T$.

2.4 Combined effects of linear and non linear distorsions. Optical solitons

Assuming a typical impulse shape, increasing for negative values of t, reaching its maximum value for t=0 and decreasing for t positive, relation (4) shows that the frequency modulation induced by non linear effects causes a frequency increase for t positive (blue shift of the rear edge) and a frequency decrease for t negative (red shift of the front edge). If β_2 is positive, the group delay increases with frequency, the front edge (resp. the rear edge) propagates more (rep. less) rapidly, and then the pulse broadens monotonically. When β_2 is negative, linear and non linear distortions induce opposite effects and the impulse narrows before, and then broadens monotonically again.

Exact resolution of the non linear Schrödinger equation which governs propagation through the fiber shows that for specific conditions, linear and non linear effects can mutually compensate along the fiber, leading to a particular impulse shape which remains undistorted while propagating in a lossless fiber : this is the optical soliton, which offers potential applications in very long distance and high capacity communication systems [*Hasegawa*, 1969 ; *Georges*, 1994].

Complex envelope of the first order soliton is expressed as :

$$E(T,Z) = \sqrt{\frac{2c}{\varepsilon_0 \omega_s n_2 Z_c}} \operatorname{sech}\left(\frac{T}{\tau_c}\right) \exp\left(i\frac{Z}{2Z_c}\right) \tag{8}$$

T is the time relative to the impulse center, Z_c and τ_c are characteristic length and time, ω_s is the angular frequency. Soliton stability being due to the exact compensation of linear and non linear effects, soliton power P_s and duration τ cannot be chosen independently. They are related by :

$$P_s = \frac{\lambda A_{eff}}{4\pi n_2 Z_c} = \frac{\lambda^3 A_{eff} D}{4\pi^2 c\, n_2 \tau_c^2} = 0{,}776\frac{\lambda^3 A_{eff} D}{\pi^2 c\, n_2 \tau^2}$$

As an example, at 1.55 µm, with usual fiber characteristics, power and characteristic distance are given by :

$$P_s = 1{,}5\,\frac{D}{\tau^2} \qquad Z_c = 0{,}25\,\frac{\tau^2}{D}, \tag{9}$$

with D in ps/(nm.km), τ in ps, P_s in mW and Z_c in km. For dispersion shifted fiber (D=1 ps/(nm.km)) and 20 ps solitons, relations give a power of 4 mW and a characteristic distance of 100 km. For a given duration, required power increases with dispersion parameter D and effective area A_{eff}, because a stronger non linear effect is needed to compensate for a larger linear distortion.

Theoretically solitons exist only on lossless fibers, which do not exist, and in the absence of any perturbation, for instance optical amplifiers, which are in fact used in systems to compensate for losses. Fortunately, it can be shown for example that induced perturbation remains tolerable if amplifier spacing is smaller than Z_c, and this conclusion remains valid for other types of periodic perturbations : when their spatial period is small compared to Z_c, soliton is only affected by the averaged perturbation [*Georges*, 1994].

3 Sources and modulators

3.1 Sources

Although electroluminescent diodes (LED) emitting at 0.8 µm were used in the first telecommunication systems on multimode fibers, semiconductor lasers, also called laser diodes (DL) are now the universal light source for high capacity optical communication systems. The basic principle of light emission is very similar to that of any microwave oscillator, even if physical phenomena involved are different : the injection of current through a PN junction provides gain in the material, which is placed inside an electromagnetic cavity constituted by the cristal itself ended by (cleaved) semi reflecting facets. When the injection current exceeds some threshold, i.e when the gain exceeds the losses due to the material itself and the reflection on the facets, system begins to oscillate [*Agrawal*, 1986].

Due to the cavity length (typically 300 µm) compared to the wavelength, mode spacing is small compared to the gain bandwidth of the material. Although the gain curve is usually regarded as being homogeneously broadened (which theoretically allows only one mode to oscillate), spontaneous emission compensates for the gain-loss difference of modes near the oscillating one and permits multimode operation. This is a serious drawback for communication systems.

In order to select only one mode, the basic structure depicted before, i.e the Fabry Perot laser, is now replaced by Distributed Feedback (DFB) or Distributed Bragg Reflector (DBR) structures. Instead of being obtained by semi reflecting facets, the cavity is now realized by inserting a grating inside the material itself, which allows only one mode near its own Bragg frequency to oscillate. Figure 2 depicts a standard Fabry Perot laser and a DFB structure.

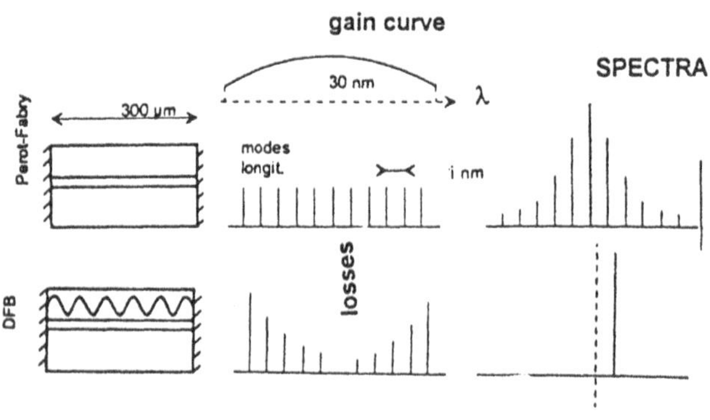

Figure 2 : Perot Fabry and Distributed Feedback laser.

As any electromagnetic oscillator, a laser does not emit a perfectly sinusoïdal wave. It is affected by amplitude and phase noise which contribute to degrade the performance of optical communication systems [*Petermann*, 1988]. Amplitude noise is usually characterized by the relative intensity noise (RIN), defined as the ratio of the power spectral density of the power fluctuation normalized by the mean emitted power [*Joindot*, 1992]. Then if the emitted power is written as $<P>+\Delta P$ where P is the average power and ΔP its random fluctuation, $rin(f)$ is defined as

$$rin\,(f) = \frac{1}{<P>^2} \int <\Delta P(t)\,\Delta P(t-\tau)> exp\,(-2i\pi f\tau) d\tau, \tag{10}$$

and practically, the parametre used is *RIN(f)* defined as 10log (*rin(f)*), expressed in dB.s or dB/Hz. The best lasers avalaible today exhibit *RIN* values better than of -155 dB/Hz and amplitude fluctuations are not actually a limiting factor for the high capacity digital optical systems. But *RIN* is a very critical parameter for video broadcasting systems using a multiplex of analog modulated carriers.

Frequency fluctuations can be considered as white, which leads to the conclusion that phase fluctuations are a diffusion process and the output spectrum of the laser is lorentzian. Phase fluctuations are not critical for the systems under consideration, as it will be shown later.

3.2 Modulators

The capacity of any oscillator to be used as a source in a communication system is strongly related to its modulation capability. All the digital optical systems use intensity modulation, obtained either by modulating the injected current (direct modulation), or by using an external modulator.

3.2.1 Direct modulation

Systems using the first technique suffer limitations due to the chirp, i.e the frequency modulation associated with amplitude modulation. The physical reason is that any modulation of carrier density in the cavity causes refractive index and therefore frequency fluctuations. It can be shown that if $P(t)$ is the time dependent emitted optical power, the difference between the instantaneous frequency $\nu(t)$ and its average value $<\nu>$ is given by :

$$\nu(t) - <\nu> = -\frac{\alpha_H}{4\pi}\left[\frac{\mathrm{d}\ln P(t)}{\mathrm{d}t} + k_0 P(t)\right] \tag{11}$$

The first term is called transient chirp and is dominant at high modulation frequency compared to the second one called adiabatic chirp. α_H is a parameter characterizing the degree of chirp of the source and called line enhancement factor, or chirp parameter. Its typical value is -5 or -6 for usual lasers.

Chirp combines with chromatic dispersion, as in the non linear case self phase modulation does. In the case of a gaussian impulse, analytical computation can be performed again. The impulse remains gaussian, but its temporal width changes and is given at distance L by :

$$\theta(x) = \theta_0\sqrt{(1 - \alpha_H x\,\mathrm{sign}\beta_2) + x^2} \qquad (12a)$$

where θ_0 and x have been defined in (3). Relation (8) shows that two cases can be distinguished, according to the sign of the product $\alpha_H\ \beta_2$. When it is negative the impulse broadens monotonically, while when it is positive, it suffers a narrowing, reaches its minimum width and then broadens again. Due to the fact that α_H is negative for a laser, these two situations correspond respectively to the normal and anomalous regime defined in section 2.1. Using (8), it is possible to derive the length $L(\alpha_H)$ corresponding to a broadening by a factor $\sqrt{2}$ and to show that :

$$\frac{L(\alpha_H)}{L_D} = \frac{\alpha_H\,\mathrm{sign}\,\beta_2 + \sqrt{1 + 2\alpha_H^2}}{1 + \alpha_H^2} \qquad (12b)$$

where the dispersion L_D has been defined by (3). With a typical chirp parameter of -5, $L(\alpha_H)$ is only 8% (resp. 47%) of the dispersion length for chirpless impulses in normal (resp. anomalous) propagation conditions and these values give an idea of the transmission length reduction due to chirp.

3.2.2 External modulation

The first type of modulator is based on electrooptical effects occurring in some materials like lithium niobate ($LiNb0_3$) : index variations obtained by applying an external voltage result into phase modulation of the optical wave propagating through the material. Phase modulation can be converted into amplitude modulation by inserting the phase modulator in one arm of a Mach Zehnder interferometer, or by using a modulator in each arm (push pull). Bandwidth can reach 20 GHz, with typical values of voltage V_π needed to cause a phase shift of π typically equal to 10 V.

Electroabsorption (EA) modulators [*Devaux*, 1993] use the change of the absorption spectrum of a semiconductor submitted to an external electrical field. Increasing the applied voltage shifts the absorption edge towards higher wavelengths and then increases light attenuation. Typical characteristics are an extinction ratio of 10 dB/V for a standard device (250 µm long) and an input power around 10 mW. A major interest of EA modulators is the possibility of integrating laser and modulator on the same substrate and obtaining a device called "laser modulator", which is very promising for future high capacity optical communication systems.

The advantage of external modulation is a very small chirp compared to direct modulation : the typical absolute maximum value of chirp parameter in this case is around 1. It is even possible to adjust the chirp in order to compensate for frequency modulation due to propagation, and then increase the transmission length. Modulator characteristics and especially chirp parameter are then crucial.

4 Detection

4.1 Basic principles of direct detection

Optical power fluctuations carrying the information to be transmitted are converted into electrical current in the photodetector, which can be a PIN photodiode or an avalanche photodiode (APD) : incident photons illuminating the photoconductor junction create photoelectrons generated according to a random Poisson process [*Personick*, 1973]. In the case of the PIN photodiode, the mean value of the photocurrent I is given by :

$$I = \frac{e\eta}{h\nu_s} P = SP, \tag{13}$$

where e is the electron charge, h the Planck's constant, ν_s the optical frequency and η the quantum efficiency, which can be viewed as the percentage of incident photons creating effectively a photoelectron, and P the incident optical power. Relation (13) shows that the detection process is quadratic, i.e photocurrent is proportional to the square of the incident electromagnetic field. Typical values of η and S at 1.55 µm are around 0.7 and 0.8 A/W. Shot noise inherently related with the corpuscular structure of light is also generated in the detection process. Its two-sided power spectral density is given by the classical Schottky's formula :

$$\gamma_g(f) = eI = \frac{e^2\eta}{h\nu_s} P. \tag{14}$$

Avalanche photodiodes differ from PIN photodiodes by the fact that each incident photon gives rise not only to one but more generally to a random number m of photoelectrons [*Webb*, 1974]. The photocurrent mean value increases compared to the case of PIN photodiode but also does the noise spectral density. They are respectively given by :

$$I = S\mathbf{E}(m)\,P \qquad \gamma_g(f) = e\mathbf{E}(m^2)SP, \tag{15}$$

where $\mathbf{E}(.)$ denotes the expected value.

Photodetector is followed by an electronic amplifier which adds gaussian thermal noise current, usually characterized by its power spectral density in A^2/Hz, to the received signal. At the output of the electronic amplifier, a decision device decides what data was received by comparing the signal at the sampling time to a threshold. Shot and thermal noise contribute to the error probability, but usual photoreceivers used in digital communications systems are limited by additive thermal noise more than shot noise.

In the ideal case of a noiseless electronic amplifier, the signal to noise ratio ρ, i.e the ratio of the signal power to the noise power in a given bandwidth B, can be expressed as :

$$\rho = \frac{\eta P}{h\nu_s B}. \tag{16}$$

Although noise properties are quite different (shot noise is not an additive noise), expression (16) is formally identical to what is obtained in the radio domain, $k\Theta$, where k denotes the Boltzmann's constant and Θ the noise temperature being replaced by h and ν_s, while $1/\eta$ can be considered as a noise figure. Assuming η equal to 1, (16) represents an intrinsic signal to noise ratio, due to the (random) corpuscular nature of the electromagnetic wave, made of photons emitted according to a Poisson process.

4.2 The quantum limit

Assuming an emitter sending either a light pulse of duration T, power P, or no light according to the binary data to be transmitted, the optimum receiver is then a photon counter and, using the Poisson distribution, the error probability is expressed as :

$$P_{eq} = \frac{1}{2}\exp\left[-\frac{PT}{h\nu_s}\right] = \frac{1}{2}\exp\left[-\frac{P}{h\nu_s B}\right] = \frac{1}{2}\exp\left[-\frac{2P_m}{h\nu_s B}\right] = \frac{1}{2}\exp\left[-2\rho_q\right], \tag{17}$$

where P_m is the average power. This ideal situation when only shot noise is present is referred as the quantum limit and is used as a reference in optical communications. Relation (14) shows that an error probability of 10^{-9} requires ρ_q to be equal to 10, which in other terms means that 10 photons must be transmitted per transmitted bit.

4.3 Computation of the bit error probability for a direct detection receiver

Assuming the case of binary equally distributed data, the average value of the signal at the photodetector output at sampling time can take on two values I_0 and I_1 according to the transmitted data. Noise contributions are respectively the signal dependent and non gaussian shot noise and signal independent and gaussian noise. Considering the total noise as gaussian with a signal dependent variance σ^2_0 or σ^2_1, and assuming an optimal decision threshold, the error probability can be expressed by the following relationship :

$$P_e = \frac{1}{2}\,\mathrm{erfc}\left[\frac{I_1 - I_0}{(\sigma_1+\sigma_0)\sqrt{2}}\right] = \frac{1}{2}\,\mathrm{erfc}\left[\frac{Q}{\sqrt{2}}\right] \qquad (18)$$

$$\sigma_k^2 = 2eI_kB + \sigma_{th}^2 \quad k = 0,1,$$

where σ^2_{th} is the thermal noise power and erfc(.) the complementary error function, widely used in digital communications and given by :

$$\mathrm{erfc}\ X = \frac{2}{\sqrt{\pi}} \int_X^{+\infty} \exp(-t^2)\,\mathrm{d}t \approx \frac{\exp(-X^2)}{X\sqrt{\pi}} \text{ for } X \geq 3. \qquad (19)$$

Figure 3 shows the error probability in logarithmic scale versus $10\log Q^2$.

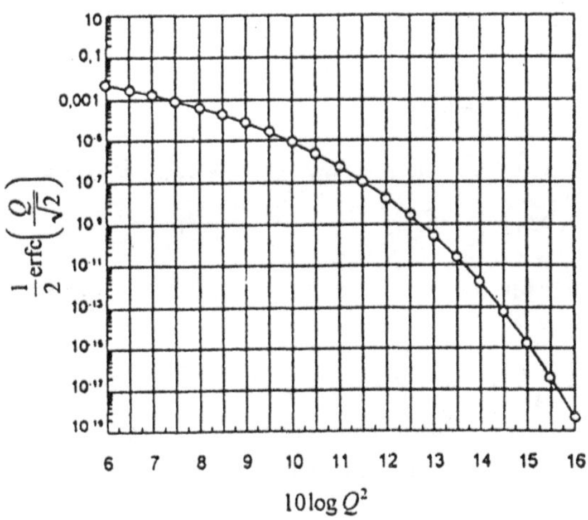

Figure 3 : Error probability versus Q^2 (in dB).

In order to ensure an error probability less than 10^{-9}, considered as a reference value for telecommunication systems, Q must exceed 6. Note that, due to the signal to noise ratio encountered in digital communication systems, approached expression given in (19) always holds.

Usual receivers performance is limited by thermal noise and the required received power to ensure a given error probability exceeds commonly by 15 or 20 dB the quantum limit. In order to improve this performance, coherent reception appeared as a promising technique. An important research effort was devoted to it up to the beginning of the nineties, before optical fibre amplifiers became available. We will briefly recall the basic principles of coherent receivers. A lot of BER computation methods were developed [*Dogliotti*, 1979]

4.4 Coherent reception

As the first radio receivers used direct detection through a crystal, heterodyne receiving technique allowed to increase dramatically the sensitivity of the receivers : the idea was to let the incoming modulated wave at angular frequency ω_s beat in a non linear device called mixer with a sinusoidal signal at angular frequency ω_l produced by the local oscillator. The beat resulted in a signal at the intermediate angular frequency ω_i carrying the modulation of the received signal. Moreover, heterodyning allows all types of modulation to be used, while frequency or phase modulation cannot be detected by a direct receiver which only detects power fluctuations [*Salz*, 1985, *Joindot*, 1986].

These receivers very commonly used today in radio can be theoretically implemented in optics. Assuming for example a phase modulated signal which can be written as :

$$u(t) = \sqrt{2P_s}\cos\left(\omega_s t + \theta_0 + \Phi\right) \tag{20}$$

where Φ equals 0 or π according to the transmitted data and a local oscillator wave :

$$u_L(t) = \sqrt{2P_L}\cos\left(\omega_L t + \theta_0\right) \tag{21}$$

the resulting beat signal (photocurrent) at the photodetector output has an average value I given by :

$$u_I(t) = 2\frac{e\eta}{h\nu_s}\sqrt{P_s P_L}\cos\left(\omega_I t + \Phi\right) \tag{22}$$

As far as noise is concerned, due to the fact that power P_L is always large compared to the received signal power, shot noise associated with the local oscillator wave is dominant. It is white with two sided power spectral density (cf. relation (14) :

$$\gamma_I(f) = \frac{e^2\eta}{h\nu_s}P_L \tag{23}$$

The carrier to noise ratio ρ, defined as the ratio of the intermediate frequency signal power to the noise power contained in a bandwidth B=1/T determines the error probability. It is given using (22) and (23) by :

$$\rho = \frac{\eta P_s T}{h\nu_s} = \frac{\eta P_s}{h\nu_s D}, \tag{24}$$

and the bit error rate is given by :

$$P_e = \frac{1}{2}\operatorname{erfc}\rho = \frac{1}{2}\operatorname{erfc}\sqrt{\frac{\eta P_s}{h\nu_s D}} \approx \frac{\exp(-\rho^2)}{\rho\sqrt{\pi}} \tag{25}$$

This relation shows (assuming a perfect detector, η=1) that the coherent receiver can approach the quantum limit and then provide a significant sensitivity improvement compared to direct detection receivers by eliminating the influence of thermal noise. But practical implementation is not easy. For example, oscillator phase noise is a serious problem, much more difficult to solve than in microwaves, because the linewidth is much larger compared to the bandwidth of the modulating signal. After a lot of efforts had been made to overcome these drawbacks, optical fibre amplifiers associated with direct detection receivers provided the same sensitivity improvement without the aforementioned problems and research about coherent techniques decreased very strongly and was abandoned in most laboratories.

5 Optical amplification

The first optical amplifiers to be studied were semiconductor amplifiers (SOA), which use the same physical phenomena as lasers : pumping in a material through electrical carriers injection (i.e through an electrical current providing the external energy) causes a population inversion. Electrons on the upper overoccupied level fall down onto the fundamental energy level again and emit correlatively photons at a wavelength corresponding to the energy difference between the levels. Most of them add in phase with those of the incident light and contribute to its amplification : this is the stimulated emission process. But other photons are emitted incoherently with the incident light : this is the spontaneous emission. A small part of these photons travel themselves through the amplifier, are amplified and constitute at the amplifier output the amplified spontaneous emission (ASE) noise, i.e the noise generated in the amplification process.

The first theoretical works were published at the beginning of the sixties [*Shimoda*, 1957] and application to optical communication systems were proposed in the seventies [*Personick*, 1973]. This is only in the beginning of the eighties that progress about semiconductor lasers allowed to consider SOA's as practically implementable devices and a lot of investigations was devoted to them [*Yamamoto*, 1980].

In optical fiber amplifiers [*Desurvire*, 1994], the active medium is a pièce of rare earth (usually erbium) doped fiber pumped by one (or eventually two) laser diode(s) emitting at a wavelength of 980 or 1480 nm. The amplification bandwidth is around 12 nm, i.e 1500 GHz (and even around 24 nm for fluoride doped amplifiers now under study in the laboratories). The output power of these erbium doped fiber amplifiers (EDFA) can be high (up to 20 dBm), which allows to increase the transmission length, with nevertheless limits due to the counterpart of non linear effects described in section 1. Sometimes, remote pumping amplifiers are used : this is done in some undersea lightwave systems, where the active fiber itself is immerged, while pump, which is the most critical component in terms of reliability, is placed at the end of the link and feeds the amplifier itself through the fiber.

The first publications about EDFA appeared in 1987 and practically usable devices were available less than four years later. Compared to semiconductor amplifiers, fiber amplifiers are easier to implement in practical systems, and present the advantage to be nearly polarization insensitive : they can be used as power emitting amplifiers (boosters), preamplifiers at the receiving end, or in line amplifiers. In this last case they can replace electronic regenerators, as for instance in the last generation of transoceanic undersea lightwave systems. Nevertheless, the counterpart is noise, linear and non linear distortions.

It is important to remark that, although they do not appear as the most promising candidate for amplification in optical communication systems, SOA's exhibit very interesting non linear properties, which make them key devices for optical signal processing (reshaping, sampling...) which will be more and more used in future optical networks.

5.1 Noise characteristics of optical amplifiers

An optical amplifier adds then its own noise to the amplified signal, as any electrical amplifier also does. Optical electromagnetic field associated to ASE noise can be modelized as a white gaussian process and the optical noise power spectral density (p.s.d) per mode at the output of the amplifier is writtten as :

$$\gamma_{esa}(\nu) = n_{sp}(G - 1)h\nu_s \qquad (26)$$

where G is the power gain and n_{sp} the spontaneous emission factor characterizing the amplifier. In the case of a semiconductor amplifier, (26) gives the noise p.s.d while, in the case of a fiber amplifier, the total power in a band B is $2\gamma_{esa}B$ because each mode is degenerated according to the two possible orthogonal states of polarization. As in the case of an electronic amplifier, the amount of noise due to the amplifier can be characterized by the noise figure. ASE noise can be considered as due to a noise source with a p.s.d γ_{ase}/G at the input of a noiseless ideal amplifier. Noise at the amplifier output can be viewed as the sum of shot noise and ASE noise and the signal to noise ratio can be written as :

$$\rho_{out} = \frac{GP_s}{(h\nu_s + n_{sp}(G-1)h\nu_s)B} = \frac{P_s}{F_a h\nu_s B} = \frac{\rho_{in}}{F_a} \qquad (27)$$

where ρ_{in} is the "intrinsic signal to noise ratio" at the amplifier imput (cf. (16)) and F_a the amplifier noise figure expressed as :

$$F_a = n_{sp}\frac{G-1}{G} + \frac{1}{G} \qquad (28)$$

If G is large, noise figure equals practically n_{sp}. It is also important to remark that losses in connections to the input fiber must be taken into account in a practical amplifier. If C_1 is the transmission factor of the input connector, measured noise figure is :

$$F_a = \frac{n_{sp}}{C_1}\frac{G_{ff}-1}{G_{ff}} + \frac{1}{G_{ff}} \qquad (29)$$

where G_{ff} is the fiber to fiber gain, i.e including losses in input and output connectors. Figure 4 depicts the gain (in dB) and noise parameter n_{sp} of a typical EDFA versus the imput signal power (in dBm).

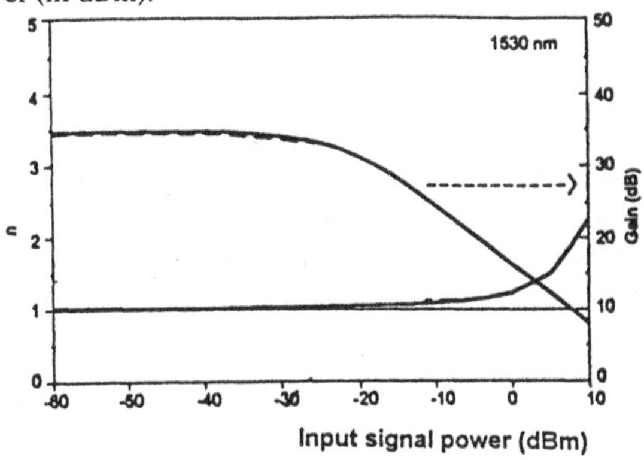

Figure 4 : Gain and noise parameter of a typical erbium doped fiber amplifier.

5.2 Sensitivity improvement brought by an optical amplifier

Let us assume an optical receiver with a preamplifier before the photodetector. Due to the fact that signal power is always large enough compared to noise power, the main noise contribution results from the beat between the optical signal and ASE noise (called ssp beat) in the quadratic detector. Self beating of ASE noise (called spsp beat) must also in some cases be taken into account. Shot noise is negligible. Assuming that only ssp beat is significant, the electrical to noise ratio at the output of the electronic amplifier can be written as :

$$\rho = \frac{S^2(GP_s^2)}{2S^2 n_{sp} G(G-1)P_s h\nu_s B + \sigma_{th}^2} = \frac{P_s}{h\nu_s B}\left(2n_{sp}\frac{G-1}{G} + \frac{\sigma_{th}^2}{S^2 G^2 P_s h\nu_s B}\right)^{-1} \quad (30)$$

Relation (30) shows that, if G is large enough, ρ is approximately given by :

$$\rho = \frac{P_s}{2n_{sp} h\nu_s B} \quad (31)$$

The influence of thermal noise is completely eliminated because, similarly to what happens in radio receivers, the noise figure of the electronic preamplifier is completely masked by the high gain of the first stage. In this asymptotic situation, receiver sensitivity is only limited by amplifier ASE noise. Considering an usual value of n_{sp} equal to 1.5, (31) shows that the degradation compared to the ideal case when only shot noise exists is 4.7 dB, which proves the sensitivity improvement due to the optical preamplifier. Due to the fact that direct detection is mostly used in optical communication systems, noise figure is often but not always defined through (31), i.e as $2n_{sp}/C_1$, at large gain, twice the value given by (29).

5.3 Amplified systems

As already mentioned, the insertion of in line amplifiers along the fiber allows to compensate for fiber attenuation and then increase the transmission length. Moreover, replacement of electronic regenerators by optical amplifiers provides (at least theoretically) transparency, i.e there is no more clock frequency dependent electronics and equipement simplification, since an optical fiber amplifier is a priori simpler and more reliable than the association of a photodetector, a laser, and electronic circuits for amplification, clock recovery.... Nevertheless, the counterpart of in line amplification is noise and distortion accumulation along the transmission line, and in line amplification can be viewed in some ways as a an evolution back towards analog transmission, which digital technique replaced in the lasr decades.

Assuming N amplifiers with a gain G compensating exactly for the fiber loss between two successive amplifiers, the total noise p.s.d at the receiver input is given by :

$$\gamma_{ase}(\nu) = n_{sp} N(G-1)h\nu_s = n_{sp}\frac{L}{L_a}(G-1)h\nu_s = n_{sp}\,\alpha L\frac{(G-1)}{\ln G}h\nu_s, \quad (32)$$

where L is the total length of the link, L_a the amplification span, and the attenuation factor related to G by $G=\exp(\alpha L_a)$. Relation (29) shows that noise contributions of all the amplifiers add exactly. In order to keep the same signal to noise ratio, for a given L_a, the output power (in dBm) must then be increased by $10\log N$. But the non linear distortion will increase correlatively, which means that the length cannot be increased indefinitely. (29) shows also that, for a given L, noise power is minimum when G approaches 1, i.e when amplifier spacing tends to 0 : there is obviously a practical limit for system cost reasons.

6 Wavelength division multiplexing (WDM)

In order to take advantage from the enormous bandwidth provided by optical fiber and optical amplifiers (typically 12 nm, i.e 1500 GHz, or even 24 nm, i.e around 3000 GHz with wideband amplifiers), Wavelength Division Multiplexing (WDM) appears as an attractive solution. The basic idea is very well known in the radio community : it consists in transmitting several channels within the available bandwidth. A main advantage is the possibility of transmitting a very high bitrate which could not be carried on one carrier on the same distance because of limitation due to chromatic dispersion and polarization mode dispersion for instance and also because of unavailability of fast enough electronic circuits. Today, systems with 4 or 8 channels at 2.5 Gbit/s are about to be commercially available, although not yet installed in the telecommunication networks, but will certainly be widely deployed in the next years. In the future, systems with more channels or a higher bitrate per channel (10 Gbit/s or more) can be envisaged. A lot of results have been obtained in many laboratories in the world, but research must yet be carried on in order to determine WDM capacity limits.Table 1 shows some results obtained in different laboratories, with bitrates per channel greater than 10 Gbit/s.

Laboratory	WDM Multiplex	Length (km)	Comments
AT&T (1994)	8 x 20 Gbit/s	300	DSF
AT&T (1994)	17 x 20 Gbit/s	150	DSF
CNET (1995)	8 x 20 Gbit/s	300	Fluoride amplifiers
AT&T (1995)	8 x 20 Gbit/s	232	DCF
NTT (1995)	16 x 10 Gbit/s	1000	SSMF and DCF

Table 1 : Some recent results in WDM technique

Another announced advantage of this technique was flexibility, because gradual increase of capacity of a communication system by progressively adding new channels could appear very attractive. In fact, this is not so simple, because new physical phenomena related with the simultaneous transmission of several optical carriers will also induce some degradations as a counterpart. In particular, cross phase modulation and Four Wave Mixing already described in section 2 are of primary importance. Figure 5 shows a WDM spectrum with 8 channels separated by 4 nm (roughly 500 GHz) after transmission through 3 optical amplifiers.

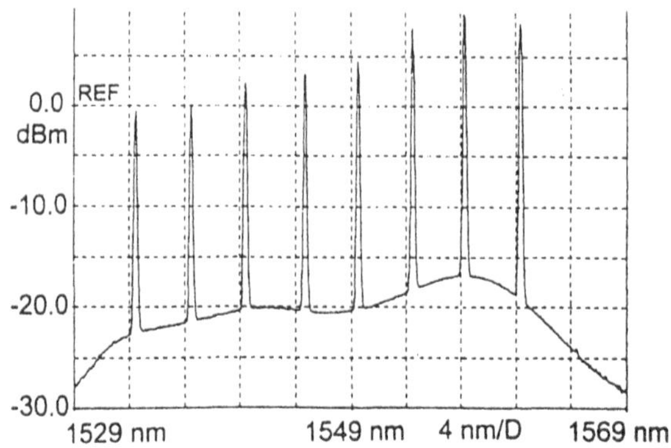

Figure 5 : Example of a WDM spectrum.

Although mathematical models exist, especially with the simplifying assumption of unmodulated channels, these phenomena are not yet completely understood : let us only recall that mismatching of channels due to high dispersion reduces the influence of these non linear effects. Channels near the zero dispersion wavelength λ_0 are then particularly affected : λ_0 is a critical parameter and more generally determination of the characteristics of fibers for future high bitrate WDM networks is a key question for telecommunication operators. Optical ampifiers do not exhibit flat gain. In amplified WDM systems, the gain is then different according to the considered channel, which means that the signal to noise ratio and consequently the bit error rate changes from one channel to another, as shown on figure 5. This is why gain equalization is a very important issue in amplified WDM systems.

7 Conclusion : trends in high capacity optical systems

High capacity optical systems are already and will be installed in terrestrial networks. After the first generation using electronic regeneration at 2.5 Gbit/s, introduction of in-line amplification and WDM at 2.5 Gbit/s per channel to increase fiber capacity will be a key issue in the immediate future. WDM systems with higher bitrate (10 Gbit/s or more) are not compatible with existing fibers and the question of the definition of new fibers to be laid in the future is a key issue for all telecommunication operators. Future of WDM versus single carrier (soliton or *NRZ*) is yet an opened question.

Optical fibers replacing the coaxial cables used before allowed a dramatic capacity increase of transoceanic links [*Taga*, 1995]. The first generation consisted in regenerated systems operating at 1.3 µm. New systems, like TAT12/13 to be laid in 1995/96, use 1.55 µm wavelength and in line amplifiers (each 40 km) instead of electronic regenerators, with the related advantages in terms of reliability and transparency, but also additional problems due to noise, linear and non linear distortions accumulation, which are more critical than in terrestrial networks because of the length, 9 000 km for a transpacific cable for instance. The next technical step will be the introduction of WDM with a bitrate of probably 5 Gbit/s per channel in the first generation. Soliton systems will be certainly used in the future : in line control techniques (sliding filtering, intensity modulation) allow to extend the transmission distance up to quasi infinite distances for bitrates up to 20 Gbit/s [*Georges*, 1995].

Another application for optical fibers is festooning, i.e connection of coastal points through an optical undersea cable, what can be very useful when geographical conditions make the installation of a terrestrial link uneasy. The objective is then to extend the transmission distance while avoiding underwater equipements and a lot of experiments have been carried out. Launched powers are high (more than 20 dBm), remote pumped post and preamplifiers and error correction coding are used.

Acknowledgments :

Irène Joindot, Michel Monerie and Patrick Vandamme are acknowledged for fruitful discussions.

References

Agrawal, G.P, N. K. Dutta [1986], *Long-Wavelength Semiconductor Lasers*, Van Nostrand Reinhold, New York.

Agrawal, G.P. [1989], *Nonlinear fiber optics*, Academic Press New York.

Desurvire, E. [1994], *Erbium Doped Fiber Amplifiers. Principles and Applications*, Wiley New York.

Devaux, F. [1993], *Fabrication, caractérisation et mise en oeuvre des modulateurs électroabsorbants en onde guidée à 1,55 µm pour les liaisons à très haut débit*, Thèse de Doctorat Paris XI.

Dogliotti, R., A.Luvison, G. Pirani [1979], "Error Probability in Optical Fiber Transmission Systems" ; *IEEE Transactions on Information Theory*, 25,170 178.

Foschini, G.P, C.D.Poole [1991], "Statistical Theory of Polarization Dispersion in Single Mode Fibers". *IEEE Journal of Lightwave Technology*, 8, 1439-1456.

Georges, T. [1994], *Contribution à l'étude des systèmes de transmission à solitons comportant des amplificateurs optiques*. Thèse de Doctorat ENST (1994).

Georges, T. [1995], "Recent advances in soliton systems" *Proc. of 21st ECOC 95 Brussel*, 312-316.

Hasegawa, A.,Y.Kodama [1969] "Signal transmission of optical solitons in monomode fibers"; *Proc. of the IEEE*, 1145-1150.

Joindot, I. [1992], *"Measurement of Relative Intensity Noise in Semiconductor Lasers"*; J. Phys III France, pp. 1591-1603.

Joindot, M. [1986], *Heterodyne Receiving Techniques in Microwaves and Optics : Comparison of some Concepts*. Digital Communications North Holland.

Mollenauer, L.F, K. Smith [1988], *"Demonstration of soliton transmission over more than 4000 km in fiber with loss periodically compensated by Raman gain"* Optics Letters,13, 675-677.

Personick, S.D, [1973] *"Receiver Design for Digital Fiber Optic Communications Systems"* B.S.T.J, 843 886.

Petermann, K. [1988] *Laser Diode Modulation and Noise*, Kluwer Academic Publishers, Dordrecht.

Poole, C.D., R.W.Tkach, A. R. Chraplyvy, D. A.Fishman [1991], "Fading in Lightwave Systems Due to Polarization Mode Dispersion"; *IEEE Photonics Technology Letters*, l.3, 68-70.

Salz, J. [1985], "Coherent Lightwave Communications"; *AT&T Technical Journal*, 64, 2153-2209.

Shimoda, K., H.Takahashi, C.H.Townes [1957], *"Fluctuations in amplification of quanta with application to Laser Amplifiers"*, J. Phys. Soc. Japan, 12, 686-700.

Sian, S.S, S.M.Webb, K.M.Guild [1995], "16 x 2,5 Gbit/s WDM Unrepeatered Transmission over 427 km" *Proc. of 21st ECOC 95 Brussels*, 975-979.

Taga, H. and al. [1995], "Recent Progress in Amplified Undersea Systems"; *IEEE Journal of Lightwave Technology*, 13, 829-840.

Thiennot, J., F. Pirio, J. B. Thomine [1993], "Optical Undersea Cable Systems Trends"; *Proceedings of the IEEE*, 81, 1610-1623.

Tkach, R.W and al. [1995], "Four Photon Mixing and High Speed WDM Systems"; *IEEE Journal of Lightwave Technology*, 13, 841-849.

Vassallo, C. [1991], *Optical Waveguide Concepts*, Elsevier, Amsterdam.

Webb, R., R. McIntyre, J. Conradi [1974], "*Properties of avalanche photodiodes*"; RCA Review, 35, 234-278.

Yamamoto, Y. [1980], "Noise and error rate performance of semiconductor laser amplifiers in PCM-IM optical transmission systems"; *IEEE Journal of Quantum Electronics*, 16, 1073-1081.

COUNTING OF SINGLE FLUX AND SINGLE CHARGE QUANTA FOR METROLOGY

Jürgen Niemeyer

1 Introduction

In the last few decades, many efforts have been made to realize the physical units by quantum devices instead of material artifacts. The latter ones imply a great number of drawbacks:

- they might be damaged by improper handling
- the realized unit depends on external parameters, often in a non-reproducible way
- intercomparisons are difficult because of transport problems
- they age
- they are available only in a central laboratory.

In contrast to this, quantum devices reduce the realization of the physical units to the use of fundamental constants which to our knowledge are independent of time and space and directly linked to the basic physical theories the consistency of which can be proved by precise determination of the fundamental constants.

Moreover a quantum "recipe" for the realization of the physical units in principle allows the calibration to be decentralized because the unit can be reproduced with fundamental accuracy in every laboratory. If the "recipe" is correctly applied, regular intercomparisons are no longer necessary.

The problem of quantum effects normally is that they are linked to a microscope scale. To measure them, the number of free parameters of the system has to be drastically reduced.

This leads in general to a reduction of system size (particle number), system temperature, and, consequently, of the measurable system signal so that it becomes impracticable for the realization of a macroscopic physical quantity. Nature fortunately produces exceptions if a physical system can be brought to coherent operation of many microscopic particles as it is for instance the case in laser systems, which provided quantum devices for the realization of time and length. Another example is today's realization of the electrical units with the help of the macroscopic quantum effects which have been detected with the aid of modern micro-lithography pattern-ing techniques and owing to the progress in low-temperature physics:

- the Josephson effect [*Josephson*, 1962]
- the quantum Hall effect [*von Klitzing et al.*, 1980; *von Klitzing*, 1986]
- single charge tunneling [*Likharev*, 1988; *Geerligs et al.*, 1990]

The Josephson effect reduces the dc voltage to the determination of the rate of transferred flux quanta $\Phi_0 = h/2e$:

$$U_j = \Phi_0 f$$

The practical unit V has been established since 1990 by definition of the Josephson constant

$$K_J = 2e/h$$

where e is the elementary charge and h Planck's constant. The value $K_{J\text{-}90}$ = 483597,9 GHz/V has been recommended by the CCE (Consultative Committee of Electricity) for maintaining the laboratory unit of volt. The value chosen is as close to the SI (Système International) value of Φ_0^{-1} as possible but it is not identical with it.

The determination of a current I by precise counting of the transfer of single elementary charge quanta e

$$I = ef$$

has been under development since 1990. The realization still suffers from the enormous experimental difficulties to be overcome by the fabrication of single electron tunneling (SET) devices.

The connection between current and voltage - Ohm's law - is established by the quantum Hall effect. The von Klitzing constant

$$R_K \equiv h/e^2$$

has been defined by the CCE in 1990 to be $R_{K\text{-}90}$ = 25812,802 Ω only for maintaining the practical laboratory unit of resistance Ω.

The relations between the electrical units according to their modern practical definition can be demonstrated in the so-called metrological triangle (Fig. 1).

The figure shows how important it would be to close the triangle experimentally by determining a current with a reproducibility of at least 10^{-8} to prove the consistancy of the metrological definitions. In addition, precise single charge measurements might be important for the redefinition of the international system of physical units and clarify discrepancies in the determination of the fine structure constant [*Taylor*, 1991; *Taylor and Cohen*, 1991].

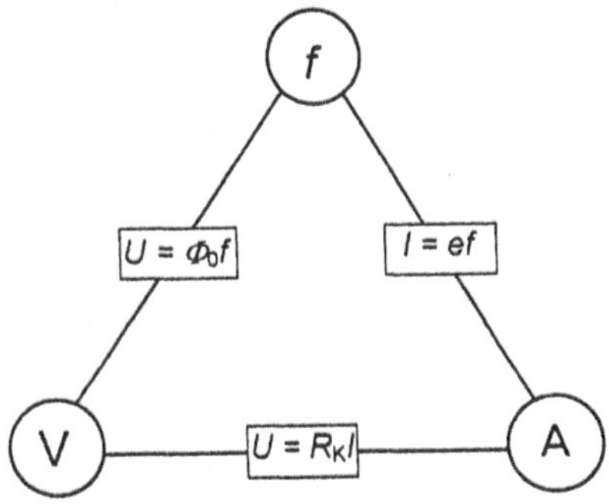

Fig. 1 The metrological triangle of the electrical units. Josephson effect and SET relate the voltage and the current via the fundamental constants Φ_0 *and e to the frequency f. Voltage and current are connected by the von Klitzing constant* $R_K = h/e^2$.

This article will describe the principal analogy between the quantum definition of the voltage and the current and the results obtained up to now by experimental realization.

2 The duality of flux quantum and charge quantum circuits

In quantum circuits physical quantities like magnetic flux F and electrical charge Q are operators rather than numerical quantities [*Widom, 1979; Leggett*, 1980; *Widom et al.*, 1982; *Widom and Clark*, 1982; *Likharev*, 1983]. Charge and flux do not commute:

$$[\Phi, Q] = i\hbar$$

This commutator can be directly derived from the well-known duality of position and momentum [*Srivasta and Widom*, 1987]. Consequently, the commutation relation is linked to the Heisenberg uncertainty relation

$$\Delta\Phi\Delta Q \geq \hbar/2$$

which expresses that flux and charge cannot be simultaneously determined with total accuracy.

At room temperature, due to the very large number of free electrons they contain, macroscopic circuits normally have an enormous number of internal electronic degrees of freedom. The wave function phases of the microscopic particles of the system are randomly distributed. To obtain macroscopic (measurable) quantities, the sum over all particles has to be formed. As a result, the phase drops out and quantum behaviour cannot be found.

The situation is completely different for systems in which, for instance, phase coherence drastically reduces the number of degrees of freedom. This is the case in superconducting circuits but also in semiconducting heterostructures (two-dimensional electron gas) or in small structures with an extremely low number of particles. In such systems, macroscopic quantum effects can be observed *[Martinis et al.*, 1987]. [*Likharev*, 1986] gives an overlook and further references.

From the uncertainty principle two different operating regimes of quantum circuits can be derived [*Clark et al.*, 1982]:

The flux regime where the magnetic flux is determined with an uncertainty smaller than the flux quantum $\Phi o = h/2e$ (h/e in the normal conducting case) and the charge regime where the charge is defined with an uncertainty better than $Q = 2e$ (e in the case of normal conductivity).

In each regime, ideal storages and transfer elements to load or unload the storage can be defined [*Prance et al.*, 1991,]; see Fig. 2:

A superconducting ring traps flux quanta nQ_0, ($n = 0, \pm 1, \pm 2, ...$) and a Josephson junction allows flux to be transferred to the ring quantum by quantum: $nQ_0, \rightleftarrows (n + 1)\, Q_0$.

An ideal, non-leaking capacitor stores charge quanta nQ_0, ($n = 0, \pm 1, \pm 2, ...$) and an extremely small tunneling element allows charge to be transferred to the capacitor quantum by quantum: $nQ_0 \rightleftarrows (n + 1)\, Q_0$.

$\Phi_0 = h/2e, Q_0 = 2e$ in the superconducting case

$\Phi_0 = h/e, Q_0 = e$ in the normal conducting case

In practice, a normal-conducting ring has to be selected extremely small so that coherence of the phase of the electron wave functions can be maintained over the ring. The transfer element must have ideally conducting electrodes for flux confinement, which are not available for experiments.

The constant rate of transferred quanta is a dc voltage $U = \Phi_0 f$ for the flux regime transfer element (Josephson junction) and a dc current $I = Q_0 f$ for the charge regime transfer tunnel junction. f is the rf-bias frequency. In the flux regime, the experimental situation is clearly dominated by macroscopic quantum effects, namely the flux quantization in a superconducting ring and the flux transfer in a Josephson junction. In contrast to this, under the charge regime, only the 2e Cooper pair transfer is determined by the coherent superconducting phase in very small junctions, where the charging energy plays an important role compared with the Josephson coupling energy.

In this case, the Cooper pairs move in a periodic potential providing voltage oscillations which are completely analogous to the current oscillations of a voltage biased Josephson junction [*Likharev and Zorin*, 1985]. These oscillations were named Bloch oscillations because the Hamiltonian of the small junctions is very similar to that of a "free" electron in the periodic potential of a semiconductor. The conditions for experimentally realizing coherent single electron transport are not yet clear but if a junction is small enough, an additional classical effect, the Coulomb blockade, will cause a non-conducting ($I = 0$) state unless the voltage exceeds a critical value $U_c = e/2C$. C is the junction capacitance. If U_c is exceeded, the junction will enter a conducting state. If in this state the junction is fed by a constant current, the charge on the electrodes will increase because $dQ/dt = I$ until $Q = e/2$. At this moment, an electron tunnels as the system energy is lowered by this. The junction charge is lowered by e and the junction will be recharged again. This periodic tunneling of single electrons results in an oscillating voltage with the frequency $f = I/e$. Phenomenologically, U_c and the voltage oscillations are analogous to the critical current and the current oscillations of a Josephson junction. Though the phenomenon of single electron tunneling is very similar to the Bloch oscillations, it should be noted that it acts on single electrons whereas the tunneling of pairs is based on the coherent operation of the pair ensemble. In that sense SET is not a macroscopic quantum effect.

In practical experiments the junctions for the transfer of pairs and single charges have to be extremely small ($C \leq 10^{-16}$ F) and highly ohmic ($R >> R_Q = h/4e^2 = 6{,}4$ kΩ). Otherwise the signal energies are lower than the fluctuation energies. The relevant energies are:

$E_{th} = kT$	thermal fluctuations
$E_Q = \hbar/\tau = \hbar/RC$	quantum fluctuations
$E_c = e^2/2C = eV_c$	charging energy of the device
$E_j = \hbar I_c/2e$	Josephson coupling energy

For the experimental realization of the diverse effects, the ratios of the competing energies must be chosen correctly [*Schmid*, 1983]:

$E_J/E_c >> 1; E_J/E_{th} >> 1; E_J/E_Q >> 1; R_Q >> R$	Josephson effect
$E_J/E_c \approx 1; E_c/E_{th} >> 1; E_c/E_Q >> 1; R_Q << R$	Bloch oscillations
$E_J/E_c >> 1; E_c/E_{th} >> 1; E_c/E_Q >> 1; R_Q << R$	SET oscillations

Table 1 shows some numerical values which result from these conditions and are to be complied with for the experimental realization. The table indicates that the flux regime is easily accessible to experiments whereas even at very low operating temperatures the proof of Bloch oscillations and SET requires extreme lithographic efforts to fabricate the required nanometer linewidth circuits.

	SET	Bloch oscillations	Josephson effect
Junction capacitance C	0,35 fF	1fF	60 pF
Junction area F	60 x 60 nm^2	100 x 100 nm^2	20 x 50 μm^2
Junction resistance R	85 kΩ	10 kΩ	5Ω
Bath temperature T	50 mK	50 mK	5Ω
Thermal fluctuation energy E_{th}	0,7 x 10^{-24} J	0,7 x 10^{-24} J	14 x 10^{-24} J
Quantum fluctuation energy E_q	3 x 10^{-24} J	1 x 10^{-23} J	0,3 x 10^{-24} J
Josephson coupling energy E_j	1,3 x 10^{-24} J	32 x 10^{-24} J	4 x 10^{-20} J
Charging energy E_c	43 x 10^{-24} J	27 x 10^{-24} J	2,7 x 10^{-28} J

Tab. 1: Numerical values for the experimental realization of quantum circuits

Including SET, at least in a phenomenological sense, the whole set of dual quantum circuits can be constructed with the basic circuit elements of Fig. 2:

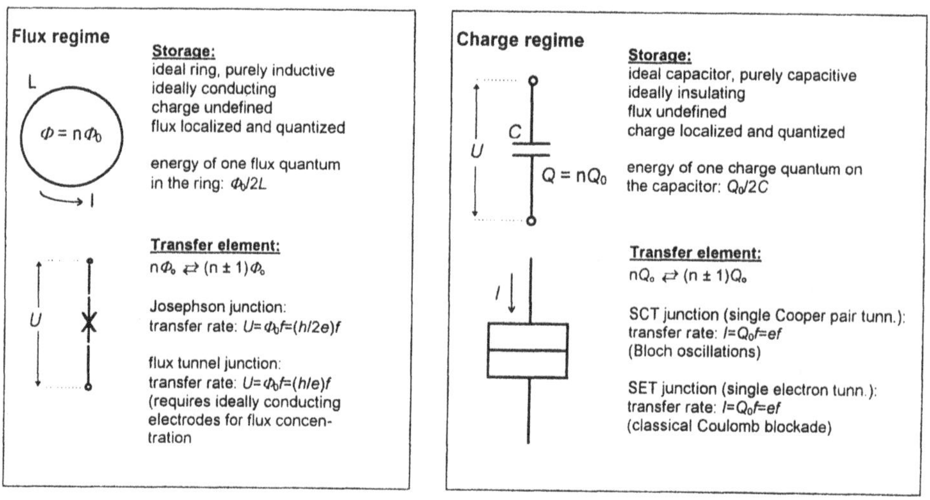

Fig. 2 Basic circuit elements for the construction of quantum circuits

The rf SQUID (a superconducting ring interrupted by a Josephson junction) corresponds to the electron box (an ideal capacitor linked to a SET junction) and the dc SQUID to a SET transistor. In analogy to the single flux quantum (SFQ) logic circuits, a complete logic circuit library can be designed with the basic SET elements. One information bit is carried by the flux quantum $\Phi_0 = \int U dt = 2$ mV ps in the first case and by one electron $e = \int Idt = 0{,}11$ μA ps in the second case. The numerical numbers show the extremely low energy consumption of both circuit types. In the case of the SET logic, the extreme smallness of the circuit elements would basically allow ultimate integration. At least the circuits in the flux regime have been realized and have commonly been used in metrology for a long time [*Likharev*, 1986]. For a few years, many SET devices - for review, see [*Devoret and Grabert*, 1992] - have been investigated and even Bloch oscillations have been demonstrated [*Kuzmin and Haviland*, 1991; *Kuzmin et al.*, 1991; *Pashkin et al.*, 1994]. While the Josephson voltage standard, which is based on the counting of the transferred flux quanta with extreme accuracy, has reached the commercial level, its analogue, the current standard on the basis of SET, is still under development and involves extraordinary experimental difficulties.

3 The Josephson voltage standard

At a finite constant dc voltage, an ac supercurrent flows through the barrier of a superconducting tunnel junction the frequency of which is directly related to the dc voltage:

$$U = \Phi_0 f_J.$$

This is caused by flux quanta traversing the junction at a certain rate, e.g. at U = 145 μV, 7×10^{10} flux quanta per second pass the junction (f_J = 70 GHz). Vice versa, if the junction is rf-biased with a frequency f, a constant voltage appears in the dc characteristic in the form of a step whose width is dependent on the rf power of the bias. Due to the strong non-linearity of the dc characteristic, this happens at higher harmonics, too:

$$U_J = n\Phi_0 f \quad (n = 0, \pm 1, \pm 2, \ldots) \qquad \text{[\textit{Shapiro}, 1963]}$$

In other words, the transfer of one single flux quantum is generated by one cycle of the rf bias.

The number of flux quanta transferred per second and, as a result, the dc voltage are determined with high precision by the stability of the frequency of the external microwave supply. In a low-damped tunnel junction with a hysteretic dc characteristic, strongly overlapping constant voltage steps can be generated in the back switching part of the dc characteristic (Fig. 3) which runs along the voltage axis at such a small distance that at low rf power, the steps cross this axis at zero current. In the case of a $Nb/Al_2O_3/Nb$ tunnel junction, the zero current steps reach a maximum of about 1 mV. At higher voltages, the steps are shifted off the voltage axis due to additional photon-induced currents across the junction. For stable step operation, the rf and dc currents have to be homogenously distributed over the junction area. This condition restricts the width and the length of the junction [*Kautz et al.*, 1987; *Niemeyer*, 1989]. As a Josephson junction is a strongly non-linear oscillator, chaotic behaviour is often dominant. To avoid chaos, the Josephson plasma frequency has to be kept smaller than the external microwave frequency [*Kautz*, 1981a; *Kautz*, 1981b; *Kautz*, 1987].

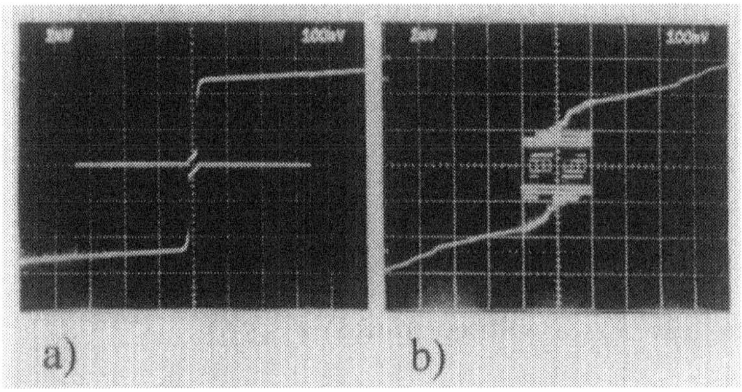

Fig. 3 Dc characteristic of an Nb/Al_2O_3/Nb tunnel junction a) without and b) with microwave radiation. rf bias frequency: 70 GHz. The constant voltage steps are crossing the current axis. Vertical axis: 1 mV/div; horizontal axis: 125 μA/div.

This condition restricts the junction critical current density and together with the junction size restrictions it leads to a limited critical current and, as a consequence, to a finite current width of the constant voltage step of less than 100 μA. But with noise-reduced equipment the voltage steps can be directly used as reference voltages in modern quantum standards. The widely overlapping zero current steps have the advantage that to increase the output voltage, series arrays of Josephson tunnel junctions can be used although the junction parameter spread is relatively large [*Levinsen et al.*, 1977]. Such arrays with more than 20 000 junctions have been fabricated in the past decade, providing reference voltages up to more than 10 V. Different technologies were tested in the past [*Niemeyer et al.*, 1985a; *Hamilton et al.*, 1985; *Niemeyer et al.*, 1985b; *Lloyd et al.*, 1987]. For practical reasons like minor aging, the Nb/Al_2O_3/Nb technology has turned out to be optimum. For proper standard operation it was important to integrate the series array into a microwave circuit which provides homogenous microwave coupling to each single junction [*Niemeyer et al.*, 1985; *Niemeyer et al.*, 1984].

To reach this goal, the series array is designed to form a superconducting microstrip line which is terminated by a lossy line to prevent standing wave patterns in the line by a reflected microwave. A fin line antenna matches the microstrip line impedance of about 5 Ω to the impedance of a 70 GHz waveguide. To achieve optimum homogeneity of the microwave coupling to the single junctions, not more than 4 000 to 5 000 junctions should be connected in a single line. Up to 5 000 junctions in a single line provide the optimum reference voltage output of 5 V or 1 mV per junction.

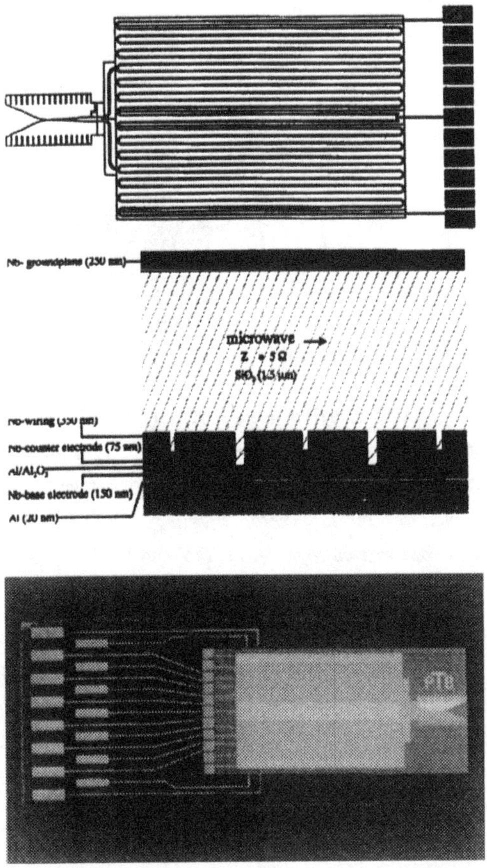

a

b

Fig. 4 a) Typical 10 V circuit design and schematic cross section of the junction stripline
b) Photograph of a 10 V standard chip bonded to a carrier.

This result shows that the microwave attenuation in the stripline is rather low. In practice, for the generation of reference voltages between 0.1 V and 3 V, circuits with two series-connected arrays of 1 500 junctions each form parallel microwave paths which are connected with the antenna via a microwave divider. A 1 V to 12 V circuit consists of four parallel microwave paths, each with 4 000 junctions. In the latter case, two quarter-wavelength transformers prevent the parallel sections from being shorted via the divider network. A typical 10 V circuit design is shown together with a photograph in Fig. 4. To generate a 10 V reference with a 70 GHz signal, a microwave power of 7 mW to 10 mW at the antenna input is required. The step amplitude at 10 V reaches 20 mA (Fig. 5) [*Müller*, 1995].

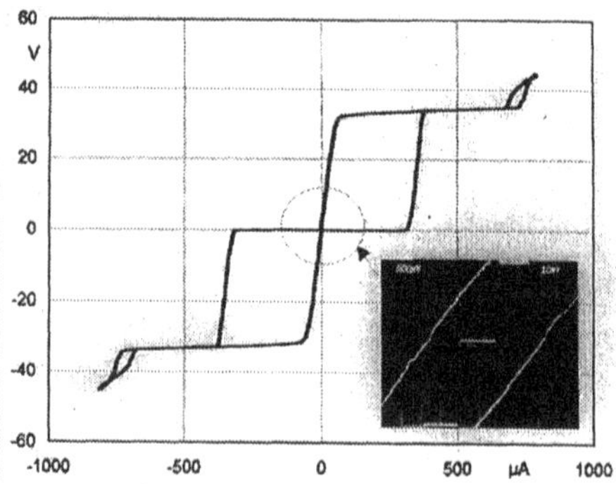

Fig. 5 Dc characteristic of a 12 000 junction series array. The inset shows how the centre part of the dc characteristic splits into 2 branches under micro-wave radiation because due to the low resolution the dc bias only follows the envelope of the constant voltage steps. For these curves the vertical resolution is 2.5 V/div and the horizontal current resolution is 10 µA/div. With the help of a multi-exposure three constant voltage steps (10 V, 0.6 V, -10 V) with the same current resolution but with a voltage resolution of 500 µV/div are inserted.

The principle accuracy of the standard is restricted by the precision of the determination of the microwave frequency apart from basic quantum deviations like macroscopic quantum tunneling [*Tsai et al.*, 1983; *Gallop*, 1991], which are practically negligible. The main source of uncertainty is the uncompensated drift of the thermal emf's in the dc leads. In practice, a reproducibility better than 5×10^{-10} V can be achieved [*Niemeyer et al.*, 1986].

The use of overlapping zero-current constant voltage steps guarantees on the one hand the possibility of adjusting reference voltages from zero to the maximum value in steps of the fundamental distance. The range between two adjacent fundamental values may be covered by tuning the external drive frequency. But on the other hand, there is a delicate stability problem related to the zero current steps: the voltage output is multivalued if the system is biased to a fixed current, e.g. $I = 0$. There is no stable voltage state between two fundamental values. This means that if the system loses the step, it might not find back to the voltage step originally adjusted and has to be readjusted again. This can be avoided to a certain degree by the use of a special bias voltage source. In particular, it is difficult to bring about fast changes of the polarity of the voltage but such changes are desirable for determining the thermal emf drift. A way out of this dilemma has been suggested in [*Benz*, 1995; *Hamilton et al.*, 1995]. A series array of highly damped, non-hysteretic Josephson junctions provides single-valued standard voltage steps at a certain current bias. The complete array is composed of a binary sequence of sections each of which is biased by a current source. A programmable multiplexer chooses the sections whose voltage output sums up to the reference voltage requested. Moreover, ac voltages with frequencies up to the MHz range can be synthesized with the fundamental accuracy of the dc standard. First versions have reached an output reference voltage of 250 mV. A sophisticated technology is required to realize this standard type: Depending on the drive frequency, a 1 V circuit is composed of 7 000 to 50 000 junctions and a 10 V series array even needs 10 times this number with a parameter spread of less than 20% to 30%. No junction failure is allowed, even a superconductively shorted junction is prohibited in the programmable version, whereas the traditional standard is not sensitive to shorts and tolerates a significantly larger junction parameter spread.

In spite of the difficulties described, the traditional series array standard is well established now in most of the standard laboratories all over the world and commercially available [*Niemeyer*, 1992; *Pöpel*, 1992]. Compact versions with an integrated cold semiconductor local oscillator are under development [*Hebrank et al.*, 1995].

4 The SET current standard

In analogy to the Josephson effect, constant current steps might be expected at $I = ef$ or $I = 2ef$ in the I-U characteristic if the SET or Bloch oscillations of small current-biased tunnel junctions - normal or superconducting - are phase-locked to an external rf source with the frequency f. Unfortunately, the experimental situation is much more complex because the current-biased junction is extremely difficult to realize. As a large charging energy is required - and by this, ultrasmall capacitances - it is forbidden to simply connect the small junction to the external world via leads. The lead capacitance and, at higher frequencies, the impedance of the free space

$$Z_o = \sqrt{\mu_o / \varepsilon_o} \approx 377\ \Omega$$

would simply shunt a single current-biased junction. Only at relatively low frequencies is it possible to find a high-impedance lead configuration with $Z(\omega) >> R_Q$, which allows charging effects to be obtained at a single junction. For a review and more references, see [*Devoret and Grabert*, 1992].

Therefore, a configuration which is easier to handle in experiments is a very small metallic (or semiconducting) island separated by two or more small tunnel junctions with a tunneling resistance $R_T >> R_Q$ to the electromagnetic environment (Fig. 6). When appropriate rf voltages are applied to the gates, one single charge is transferred across the central island per rf cycle in the case of the turnstile and the pump.

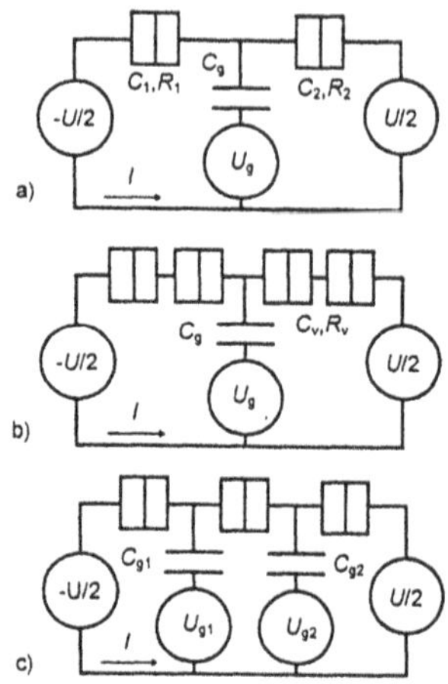

Fig. 6 Circuit diagrams of several single charge devices.
a) SET transistor; b) Turnstile device; c) Single electron pump

U : *applied voltage,*
I : *current across the tunnel junction,*
C_g, C_{gv} : *gate capacitances,*
C_v : *junction capacitances,*
R_v : *tunnel resistances,*
U_{gv} : *dc or rf gate voltages.*

The simplest circuit, the SET transistor, which was first realized by [*Kuzmin and Likharev*, 1987; *Fulton and Dolan*, 1987], forms the elementary building block of the more complex devices. The determination of its properties is therefore of importance for the operation of SET circuits. Fig. 7 shows a SET transistor of aluminium and aluminium oxide.

For preparing small tunnel junctions, a special two-layer e-beam resist mask forms small suspended bridges in the tunnel region. Two successive non-vertical evaporations, together with an oxidation process of the first layer, form small overlapping areas in the shadow of the resist bridges [*Niemeyer*, 1974; *Dolan*, 1977; *Dolan and Dunsmuir*, 1988]. Besides the possibility of manufacturing ultra-small tunnel junctions, the main advantage of the method is that the complete circuit can be produced without the vacuum cycle being interrupted.

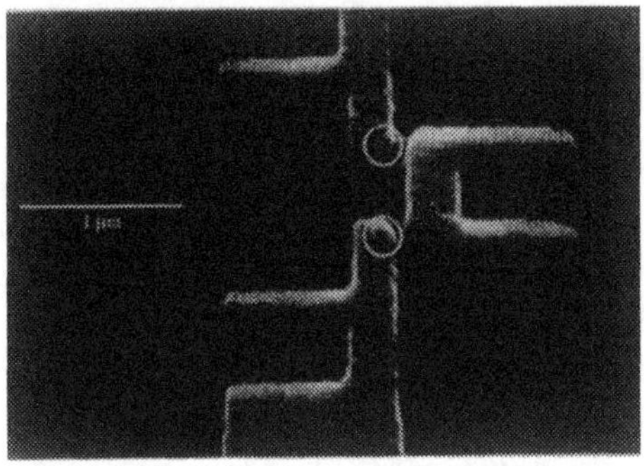

Fig. 7 SET transistor, made of Al/Al_2O_3/Al.
The circles denote the position of the tunnel junctions.

A typical dc characteristic of a voltage-biased transistor which clearly shows the Coulomb blockade and the modulation of the voltage with the gate voltage applied is shown in Fig. 8.

a)

b)

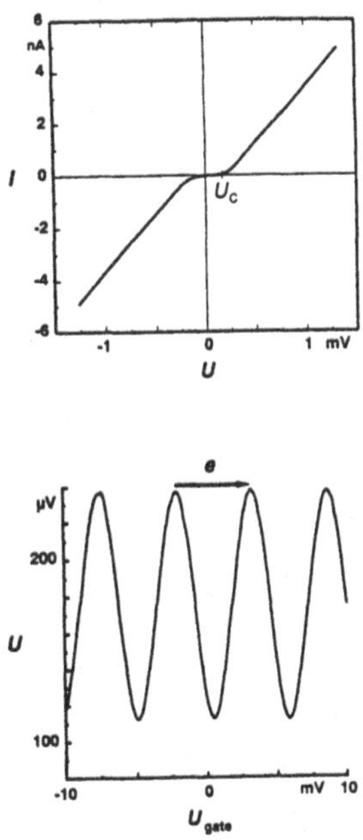

Fig. 8 a) I-U characteristic of a voltage-biased SET transistor A 187-13
b) Modulation of U with the gate U_g. Each cycle corresponds to the transfer of one electron through the circuit. A magnetic field of 1 T is applied to suppress superconductivity
($T = 50$ mK; $C_1 = C_2 = 0.4$ fE; $R_1 + R_2 = 220$ kΩ, $C_g = 0.03$ fF).

The measurements were performed in a dilution refridgerator at temperatures less than 100 mK. The apparatus is electromagnetically shielded, and all leads are equipped with special high-frequency filters. The modulation of the *I-U* characteristic with the gate voltage will be analogous to the critical current modulation of a dc SQUID, if flux quanta are entering or leaving the SQUID ring. The critical current can be considered as a flux blockade in as much as the "critical" voltage of the transistor is caused by a charge blockade, the Coulomb blockade. The charge resolution of the transistor considered as an electrometer is about 5 x 10^{-4} e/√Hz at 10 Hz, which is extremely sensitive, but because of the lead problems the electrometer function is more or less restricted to an on-chip experiment. Unfortunately, the analogue to the superconducting flux transformer does not exist in the charge regime.

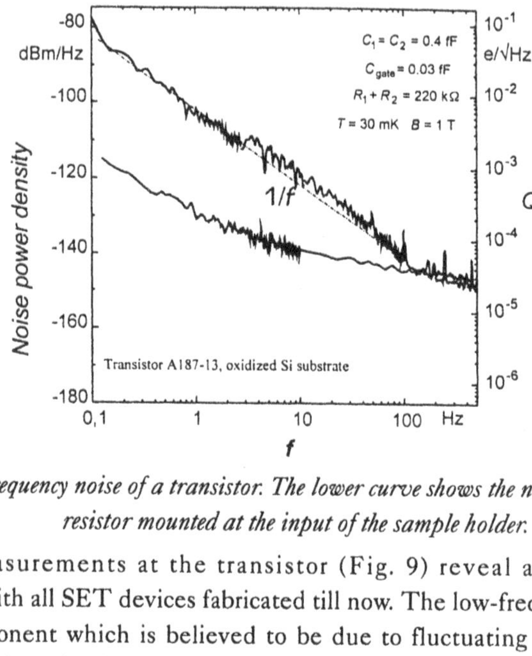

Fig. 9 Low-frequency noise of a transistor. The lower curve shows the noise figure of a 9 kΩ resistor mounted at the input of the sample holder.

Noise measurements at the transistor (Fig. 9) reveal a problem which is encountered with all SET devices fabricated till now. The low-frequency noise shows a large 1/f component which is believed to be due to fluctuating background charges trapped somewhere in the insulating regions in the vicinity of the sample or in the barrier of the junctions. Similar results have been obtained by other authors [*Zimmerli et al.*, 1992a; *Kuzmin et al.*, 1989; *Geerligs et al.*, 1990; *Zimmerli et al.*, 1992b; *Visscher et al.*, 1995; *Verbrugh et al.*, 1995; *Song et al.*, 1995]. The effect of single two-level fluctuators can be directly watched as transport voltage fluctuations (Fig. 10) the amplitude of which is on an average equal to a charge fluctuation of about 0.05 e.

The overall intensity of the charge fluctuation ranges from 10^{-3} to 10^{-4} e/√Hz at 10 Hz. The intensity of the noise does not vary with the temperature [*Verbrugh et al.*, 1995] and is dependent on the configuration of the island and the gate [*Zimmerli et al.*, 1992a; 1992b]. It could be shown that smaller islands generate less noise [*Song et al.*, 1995].

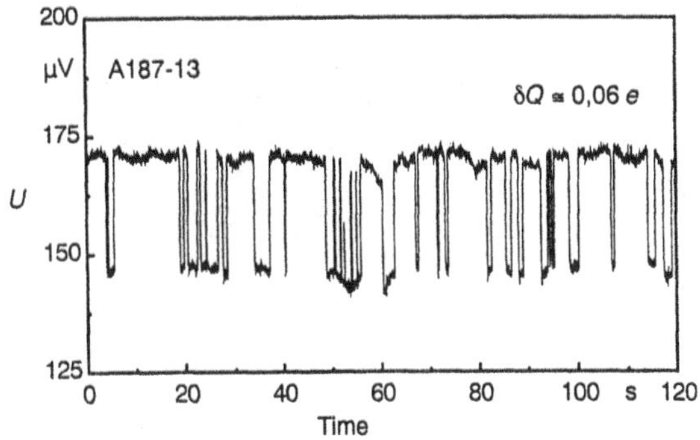

Fig. 10 Telegraph noise on the transport voltage over the time in the sensitive region of a SET transistor at I = 250 pA.

By monitoring the fluctuations of the background charges "stereoscopically" with two adjacent transistors (Fig. 11) it could be shown that a larger fraction of the charge noise probably originates from the substrate rather than from the barriers [*Martinis et al.*, 1994]. In this experiment, a correlation between the signals of the two transistors of 10% to 20% has been found (Fig. 12) [*Zorin et al.*, 1995]. Model calculations show that the relatively small correlation between the two islands can be explained best by assuming a random distribution of the background charges over a relatively large volume (several times the island width) of the substrate in the vicinity of the transistors.

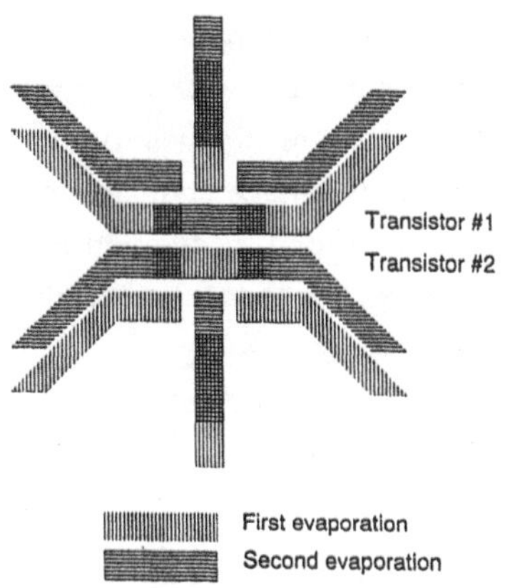

Fig. 11 Schematic figure of the "stereoscopic" transistor as a result of two oblique evaporations through a suspended bridge mask. The linewidth of the circuit is about 150 nm. The circuit has been designed and fabricated by the cryolab of Moscow State University.

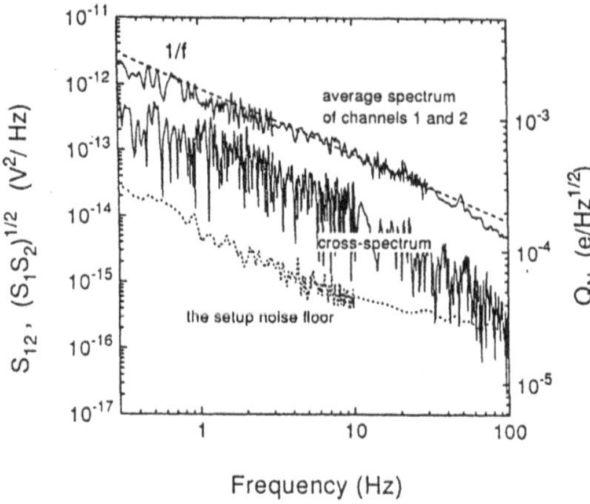

Fig. 12 The average spectrum density of the two transistors S_1 and S_2 and their cross correlation spectrum. As the measured noise power densities S_1 and S_2 lie very close together, their average value $(S_1 S_2)^{1/2}$ has been presented in this figure for better clarity.

Unfortunately, the static configuration of the background charges determine the initial state of a circuit in a more or less non-reproducible way. The accidental background charge distribution acts like a gate voltage for the island of the transistor and modulates the *I-U* characteristic in an unpredictable way. At present, this means that in all complex circuits with more than two junctions the effect of the background charge has to be compensated by gate voltages appropriately chosen for each island. Due to cross capacitances of the gates to the other islands in the close vicinity, this is very difficult in multi-junction circuits. The flux analogy is the suppression of the critical current by trapped flux inside or in the vicinity of a Josephson junction. One can easily imagine that a series array voltage standard would be unpractical if the critical currents were randomly distributed by trapped flux quanta. With a random initial state of the "critical voltages" a current standard would also be impractible.

As described above, a single SET junction cannot be used as elementary current standard due to the capacitances of the leads. Neither can the SET electrometer be used for the controlled transfer of single electrons because of the following reasons:

At zero temperature the SET transistor traps n excess electrons only close to $U = 0$ and $U_g = 2ne/2C_g$ (n =0, 1, 2, ...). n is the number of excess electrons on the island. If the applied voltage U exceeds the Coulomb blockade voltage which varies periodically between the extrema $U_c = e/(C_1 + C_2 + C_g)$ at $U_g = 2ne/2C_g$ and $Uc = 0$ at $U_g = (2n + 1)e/2C_g$, a continuous current flows across the transistor. As a result, the current is also periodic in U_g. Each period is related to an average number n of excess electrons on the island. This means that if a continuous current flows, the average number of electrons on the island will not be changed. For each electron that enters the island another one immediately leaves it. Therefore vicinity states like n = 0 and n = 1 cannot stably coexist. This prevents the controlled transfer of single charges between such states.

The simplest possible devices which transfer charges in a controlled manner are the turnstile [*Geerligs et al.*, 1990] and the electron pump [*Pothier et al.*, 1991; 1992] as shown in Fig. 6. Instead of only four or three junctions per device, more complex circuits with a larger number of junctions can be used. These circuits provide a region in the U-U_g plane where vicinity states like n = 0 and n = 1 coexist so that a controlled cycle between these states by an appropiate rf gate voltage can be experimentally realized. In both circuit types, only one electron is transferred per complete cycle of the rf voltage, providing a reference current of $I = ef$ which appears to be a constant current step in the I-U characteristic. In the case of the pump, the constant current step even crosses the zero voltage axis and only by changing the locked phase shift of the two rf-biases between $+\pi/2$ and $-\pi/2$ can the sign of the reference current be changed. This is very advantageous for precision metrology. In various experiments, the frequency of the rf drive ranges from 4 MHz to 50 MHz, resulting in reference current steps at 0.5 pA and 5 pA. At frequencies higher than 40 MHz, flat current plateaus are no longer observed due to an increasing number of cycle missing events [*Verbrugh*, 1995]. The uncertainty of the reference current experimentally achieved at this extremely low current level is at present 5×10^{-3}. This is by far not sufficient to prove the compatibility of the metrological triangle by precise current measurement in the range of 10^{-8}, to say nothing of the realization of a current standard. Besides by the cycle missing events which can be kept sufficiently small by choosing the rf frequencies smaller than 20 MHz, principal errors can be introduced only by an unwanted transfer of a single electron due to macroscopic tunneling - the so-called co-tunneling across the complete device - [*Averin et al.*, 1989], thermally activated events and events induced by high-frequency photons (> 100 GHz). Co-tunneling can be adequately reduced by the use of larger arrays, e.g. a five-junction pump or turnstiles with five junctions in each arm. Thermally activated events can be kept small enough by reducing the island size and using low rf frequencies. The high-frequency photons must be screened out by proper filters at the low temperature level of the refrigerator, e.g microfilters [*Vion et al.*, 1994] or thermo-coax cables [*Zorin*, 1995]. Considering all these measures it has been estimated in [*Pothier et al.*, 1992] that a five-junction pump with a 0.2 fF junction capacitance and a 100 kΩ junction resistance operated at 100 mK would provide a constant current step at 10 pA over a voltage range of 100 μV with an uncertainty of

10^{-8}. Similar results have been predicted by [*Averin et al.*, 1993] for the turnstile device with five junctions in each arm.

From measurements on a five-junction pump, charge counting error rates of 5 x 10^{-7} were derived [*Martinis et al.*, 1994]. This value which is higher than expected is thought to be caused by photon-induced tunneling due to imperfect filtering.

If the theoretical predictions are correct, the problem of installing a SET current scale will be "reduced" to the measurement of a 10 pA current with a relative uncertainty of 10^{-8}, which means to monitor 10^{-4} fA or about 1 electron per second on a net current of 10 pA. To solve this problem, it has been suggested to increase the current by the use of a cryogenic current comparator (in analogy to the voltage divider for scaling up voltages) with a large number of turns on the primary winding [*Gutmann and Niemeyer*, 1991; *Hartland et al.* 1991]. A number of 30 000 turns can be realized. The use of such an instrument would increase the SET output current to 0.3 mA. The best current measurements with a cryogenic current comparator have reached an absolute uncertainty of 10 fA/√Hz [*Gutmann and Niemeyer*, 1991; *Hartland*, 1992]. By averaging over an hour, 10 pA could be determined with a relative uncertainty of about 2 x 10^{-5}.

Similar results are predicted for the use of a current integrator [*Martinis et al.*, 1994; *Verbrugh et al.*, 1994; *Williams et al.*, 1992]. In this case, the small current to be determined charges a capacitor electron by electron via a pump. The electrons are counted and the capacitance of the charged capacitor is determined by comparison with a calculable standard capacitor. The voltage over the charged capacitor is measured with high precision by means of a Josephson voltage standard. In spite of the fact that the capacitance values of both the standard capacitor and the charged capacitor are very similar, the main uncertainty of this arrangement is due to the parasitic capacitances of the leads which connect the charged on-chip capacitor with the room temperature standard capacitor. As the standard capacitor Cs is basically charged directly electron by electron, the fine structure constant

$$\alpha = (\mu_0 c e^2)/(2h)$$

can be determined by relating the voltage at the standard capacitor

$$U_s = (ne)/C_s \quad (n = \text{number of transfered electrons})$$

to the Josephson voltage

$$U_j = m(h/2e)f \quad (m = \text{number of fundamental voltage steps }).$$

$U_s = U_j$ yields

$$e^2/h = mC_s f/(2n) \text{ or } \alpha = m\mu_0 c C_s/(4n).$$

c is the velocity of light, h Planck's constant and μ_0 the permeability of the vacuum. A new independent determination of the fine structure constant would be very important because there is a discrepancy in the range of 10^{-7} in the values derived from measurements of the quantum Hall effect, the proton gyromagnetic ratio in water, and the anomalous magnetic moment of the electron together with calculations according to the quantum electrodynamic theory.

All experiments so far considered show that to reach an accuracy of 10^{-8} for the current determination, an increase in the SET current by constructing a parallal array of turnstiles or pumps would be very helpful. How many elementary circuits have to be connected in parallel depends on how far the current can be increased by simpler measures such as the use of slightly higher frequencies in SET circuits or arbitrarily higher frequencies in single pair circuits with Bloch oscillations [*Kuzmin et al.*, 1991a] and if the current comparator can be further improved. A factor of 10 to 100 might be reached by such measures. This results in the requirement of connecting 100 to 10 elementary circuits in parallel. To reach this goal two problems have to be overcome:

i) The influence of the static background charges has to be reduced because a compensation of the random initial states of the "critical voltage" by ten to 100 gates seems to be unpractical. Noise measurements on SET transistor configurations as described above are an effective tool to investigate diverse substrates and circuit designs to minimize the background charges.

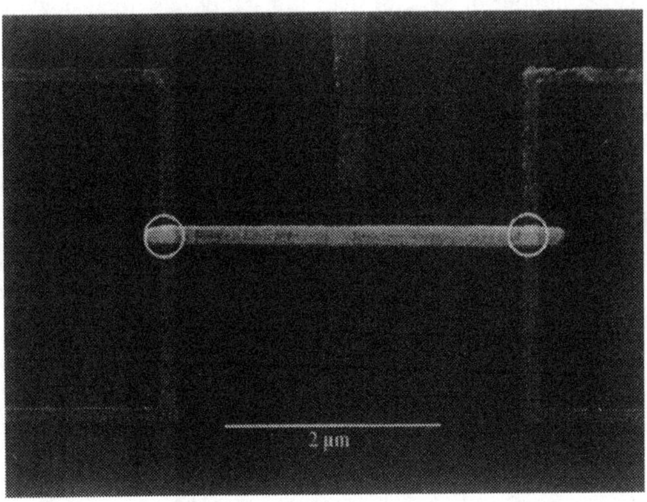

Fig. 13 Cr/CrO_x/Cr transistor. Thickness of the Cr layers: 15 nm and 50 nm. The contact wirings are resistances (> 50 kΩ).

ii) A technology has to be established which permits the fabrication of nanocircuits with reproducible properties on a linewidth scale of 50 nm to 100 nm. It should be a multilayer technolgy because the shadow evaporation technique is not versatile enough to realize complex connections between the elementary circuits. Moreover, unwanted shadows add parasitic capacitances and additional oxide layers on the islands and wires might add more background charges. A first step to develop a multilayer technology is the use of chromium as electrode material [*Pashkin at al.*, 1995]. A transistor with two $C_r/Cr_yO_x/Cr$ tunnel junctions (Fig. 13) has been made by means of two subsequently applied e- beam resist masks. One may take note of how simple the circuit looks in comparison with the transistor of Fig. 7 which has been made with the shadow evaporation method. The Cr transistor showed the expected current modulation when a gate voltage was applied. Not so clear at present is the composition of the junction barriers. On the one hand, chromium is very hard to oxidize at temperatures less than 500°C. This allows to expose the surface of the base electrode film to the ambient atmosphere for the preparation of the subsequent e-beam lithographic mask which patterns the counter electrodes. On the other hand, the resistance of chromium to oxidation favors the formation of a barrier composed of adsorbed oxygen and stable chromium oxide. A reproducible oxidation process for chromium tunnel junctions is under development at present.

If these problems are solved one should be able to increase the SET current by connecting a large number of SET devices in parallel, just in analogy to the series arrays of the Josephson voltage standard. But one should not ignore the simple fact that a single Josephson tunnel junction provides a voltage signal of about 1 mV. This was enough to realize 1V with an uncertainty of $5x10^{-8}$ by means of a voltage divider already in the early seventies, a few years only after the predictions of Josephson. In contrast to this, a single SET device provides a signal in the pA range which is by nine orders of magnitude smaller than the signal from the superconducting device, where the coherence of the phase extends over macroscopic distances. This comparison illuminates the enormous problems the experimentators face in realizing a single charge current standard.

References

Averin, D. V.; A. A. Odintsov; S. V. Vyshenskii [1993], "Ultimate Accuracy of Single-Electron DC Current Standards", *J. Appl. Phys. 73*, 1297

Averin, D. V.; Odintsov, A. A. [1989], "Macroscopic Quantum tunneling of the electric charge in small tunnel junctions", *Phys. Lett. A 140*, 251-257

Benz, S. P. [1995], "Superconductor-normal-superconductor junctions for programmable voltage standards", *Appl. Phys.* Lett. 67, 2714-2716

Clark, T. D.; R. J. Prance; T. P. Spiller; H. Prance; J. Ralph [1982], "The Flux Mode and Charge Mode Regimes in Superconducting Weak Link Circuits", *Springer Series in Electronics and Photonics*, Vol. 31, 247 - 253; "Single-Electron Tunneling and Mesoscopic Devices", Ed. H. Koch, H. Lübbig, *Springer-Verlag Berlin Heidelberg 1992*, ISBN 3-540-55132-8

Devoret, M. H.; H. Grabert [1992], "Introduction to Single Charge Tunneling", in "Single Charge Tunneling", ed. by H. Grabert, M. H. Devoret, *NATO ASI Series* Vol. 294, 1-19; Plenum Press, New York, 1992; ISBN 0-306-44229-9

Devoret, M. H.; H. Grabert [1992], "Introduction to Single Charge Tunneling" in "Single Charge Tunneling", NATO ASI Series B: Physics Vol. 294, Plenum Press New York 1992, 1-18; Ed. H. Grabert, M. Devoret, ISBN 0-306-44229-9

Dolan, G. J. [1977], "Offset masks for lift-off photoprocessing", *Appl. Phys.* Lett. 31, 337-339

Dolan, G. J.; J. H. Dunsmuir [1988], "Very Small (>,~ 20 nm) Lithographic Wires, Dots, Rings, and Tunnel Junctions", *Physica B*, 152, 7-13

Fulton, T. A.; G. J. Dolan [1987], "Observation of Single-electron charging effects in Small Tunnel Junctions", *Phys. Rev. Lett.* 59, 109-112

Gallop, J. [1991], "Macroscopic Corrections to Quantum Electrical Standards", in "Macroscopic Quantum Phenomena", ed. by T. D. Clark, H. Prance, R. J. Prance, T. P. Spiller, *World Scientific Publishing Co. Singapore*, 1991, 153-158, ISBN 981-02-0382-9

Geerligs, L. J.; V. F. Andereeg; P. A. M. Holweg; J. E. Mooij; H. Pothier; D. Esteve; D. Urbina; M. H. Devoret [1990], "Frequency-Locked Turnstile Device for Single Electrons", *Phys. Rev. Lett.* 64, 2691

Geerligs, L. J.; V. F. Andereeg; P. A. M. Holweg; J. E. Mooij; H. Pothier; D. Esteve; D. Urbina; M. H. Devoret [1990], "Frequency-Locked Turnstile Device for Single Electrons", *Phys. Rev. Lett.* 64, 2691

Geerligs, L. J.; V. F. Anderegg; J. E. Mooij [1990], "Coulomb blockade of Cooper pair tunneling", *Physica B 165-166*, 971-973

Gutmann, P.; J. Niemeyer [1991], "Design of a Current Scale Based on the Coherent Tunneling of Single Electrons Using a Superconducting Differential Flux Transformer", *PTB-Mitt.* 101, 171-174

Hamilton, C. A.; C. J. Burroughs, R. L. Kautz [1995], "Josephson D/A Converter with Fundamental Accuracy", IEEE Trans. Instr. Meas. 44, 223-224

Hamilton, C. A.; R. L. Kautz; R. Steiner; F. L. Lloyd [1985], "A Practical Josephson Voltage Standard at 1 Volt", *IEEE Elec. Dev.* Lett. 6, 623-625

Hartland, A. [1992], "The Quantum Hall Effect and Resistance Standards", Metrologia 29, 175-190

Hartland, A.; K. Jones; J. M. Williams; T. Gallagher; T. Galloway [1991], "Direct Comparison of the Quantized Hall Resistance in Gallium Arsenide and Silicon", *Phys. Rev. Lett.* 66, 969-985

Hebrank, F. X.; P. Gutmann, F. Müller, E. Vollmer, J. Niemeyer [1995], "Hybrid Integration of Microwave Oscillators with Josephson Series Arrays for the Use as Voltage Standards", *IEEE Trans. Appl. Supercond.* 5, 2911-2914

Josephson, B. D. [1962], "Possible New Effects in Superconductive Tunneling", *Phys. Lett.* 1, 251

Kautz, R. L. [1981a], "The AC effect in Hysteretic Junctions: Range and Stability of Phase Lock", *J. Appl. Phys.* 52, 3528-3541

Kautz, R. L. [1981b], "Chaotic States of rf-biased Josephson Junctions", *J. Appl. Phys.* 52, 6241-6246

Kautz, R. L. [1987], "Global Stability of phase lock near a chaotic crisis in the rf-biased Josephson junction", *J. Appl. Phys.* 62, 198-211

Kautz, R. L.; Hamilton C. A.; Lloyd, F. L. [1987], "Series-Array Josephson Voltage Standards", *IEEE Trans. Magn.* 23, 883-890

Kuzmin, L. S.; Haviland, D. B. [1991], "Observation of the Bloch Oscillations in Ultrasmall Josephson Junctions", *Phys. Rev. Lett.* 67, 2890-2893

Kuzmin, L. S.; K. K. Likharev [1987], "Direct experimental observation of discrete correlated single-electron tunneling", *JETP Ltt.* 45, 495-497

Kuzmin, L. S.; Nazarov, Yu. V.; Haviland, D. B.; Delsing, P.; Claeson, T. [1991], "Coulomb Blockade and incoherent Tunneling of Cooper Pairs in Ultrasmall Junctions Affected by Strong Quantum Fluctuations", *Phys. Rev. Lett.* 67, 1161-1164

Kuzmin, L. S.; P. Delsing; T. Claeson; K. K. Likharev [1989], "Single-electron charging effects in one-dimensional arrays of ultrasmall tunnel junctions", *Phys. Rev. Lett.* 62, 2539

Leggett, A. J. [1980], "Macroscopic Quantum Systems and the Quantum Theory of Measurement", *Suppl. Progr. Theor. Phys.* 69, 80 - 100

Levinsen, M. T.; R. Y. Chiao, M. J. Feldman, D. A. Tucker [1977], "An inverse ac Josephson effect standard", *Appl. Phys. Lett.* 31, 776-778

Likharev, K. K. [1983], "True quantum-mechanical macroscopic effects in weak superconductivity", *Usp. Fiz. Nauk* 139, 169 - 184 and Sov. Phys. Usp. 26, 87 - 95

Likharev, K. K. [1986], "Dynamics of Josephson junctions", Gordon and Breach Science Publishers, P. O. Box 786, Cooper Station New York, NY 10276, USA, ISBN-2-88124-042-9

Likharev, K. K. [1988], "Correlated discrete transfer of single electrons in ultrasmall tunnel junctions", *IBM J. Res. Dev.* 32, 144-158

Likharev, K. K.; A. B. Zorin [1985], "Theory of the Bloch-Wave Oscillations in Small Josephson Junctions", *J. Low Temp. Phys.* 59, 347 - 382

Lloyd, F. L.; C. A. Hamilton; J. A. Beall; D. Go; R. H. Ono; R. E. Harris [1987], "Josephson Array Voltage Standard at 10 V", *IEEE EDL 8*, 449-450

Martinis, J. M.; M. H. Devoret; J. Clarke [1987], "Experimental tasks for the quantum behaviour of a macroscopic degree of freedom: The phase difference across a Josephson junction", *Phys. Rev. B*, 35, 4682 - 4698

Martinis, J. M.; M. Nahum; H. D. Jensen [1994], "Metrological accuracy of the electron pump", *Phys. Rev. Lett.* 72, 904

Müller, F.; J. Kohlmann; F. X. Hebrank; T. Weimann; H. Wolf; J. Niemeyer [1995], "Performance of Josephson Array Systems related to Fabrication Techniques and Design", *IEEE Trans. Appl. Supercond.* 5, 2903-2906

Niemeyer, J. [1974], "Eine einfache Methode zur Herstellung kleinster Josephson-Elemente", *PTB-Mitt.* 84, 251-253

Niemeyer, J. [1992], "1 V and 10 V Josephson Voltage Standards" in: "Superconducting Devices and Their Applications", ed. H. Koch, H. Lübbig, Springer-Verlag, Berlin, Heidelberg 1992, ISBN 0-387-55396-7

Niemeyer, J.; J. H. Hinken; E. Vollmer; L. Grimm; W. Meier [1986], "A new Josephson voltage standard with tunnel junction series arrays", *Metrologia 22*, 213-217

Niemeyer, J.; J. H. Hinken; R. L. Kautz [1984], "Microwave-induced constant voltage steps at one volt from a series array of Josephson junctions", *Appl. Phys. Lett.* 45, 478-480

Niemeyer, J.; L. Grimm; W. Meier; J. H. Hinken; E. Vollmer [1985a], "Stable Josephson reference voltages between 0.1 and 1.3 V for high-precision voltage standards", *Appl. Phys. Lett.* 47, 1222-1223

Niemeyer, J.; Y. Sakamoto; E. Vollmer; J. H. Hinken; A. Shoji; T. Nakagawa; S. Kosaka [1985b], "Nb/Al2O3/Nb and NbN/MgO/NbN Tunnel Junctions in Large Series Arrays for Voltage Standards", *Jap. J. Appl. Phys.* 25, L343-L345

Pashkin, Yu. A.; Haviland, D. B.; Kuzmin, L. S.; Chen, C. D.; Delsing, P.; Claeson, T. [1994], "2e Periodic Modulation of the I-V-Curve of a current-biased Superconducting Transistor", *Physica B 194-196*, 1049-1050

Pashkin, Yu. A.; L. S. Kuzmin, A. N. Tavkhelidze; F.-J. Ahlers; T. Weimann; D. Quenter; J. Niemeyer [1995], "All-Chromium Single Electron Tunneling Transistor", to be published in *Appl. Phys. Lett.*

Pöpel, R. [1992], "The Josephson Effect and Voltage Standards", *Metrologia 29*, 153-174

Pothier, H.; P. Lafarge; D. Esteve; C. Urbina; M. H. Devoret [1993], "Passing Electrons Due by Due: Is a 10-8 Accuracy Acievable?" *IEEE Trans. Instr. Meas.* 42, 324-330

Pothier, H.; P. Lafarge; P. F. Orfila; C. Urbina; D. Esteve; M. H. Devoret [1991], "Single electron pump fabricated with ultrasmall normal tunnel junctions.", *Physica B 169*, 573-574

Pothier, H.; P. Lafarge; P. F. Orfila; C. Urbina; D. Esteve; M. H. Devoret [1992], "Single-Electron Pump based on Charging Effects", *Europhys. Lett. 17*, 249-254

Prance, R. J.; H. Prance; T. P. Spiller, T. D. Clark; J. Ralph; A. Clippingdale [1991], "Quantum Dualities in Superconducting Weak Link Circuits", in: "Macroscopic Quantum Phenomena", 43 - 55; World Scientific Publishing Co. Ptc. Ltd. London 1991, ed. F. D. Clark, H. Prance, R. J. Prance, T. P. Spiller; ISBN 981-02-0382-9

Schmid, A. [1983], "Diffusion and Localization in a Dissipative Quantum System", *Phys. Rev. Lett. 51*, 1506 - 1509

Shapiro, S. [1963], "Josephson Currents in Superconducting Tunneling: The Effect of Microwaves and other Observations", *Phys. Rev. Lett. 11*, 80-82

Song, D.; A. Amar; C. J. Lobb; F. C. Wellstood [1995], "Advantages of superconducting Coulomb-blockade electrometers", *IEEE Trans. Appl. Supercond. 5*, 3085

Srivasta, Y.; A. Widom [1987], "Quantum Electrodynamic Processes in Electrical Engineering Circuits", *Phys. Rep. 148,* 1 - 65

Taylor, B.N. [1991] "The Possible Role of the Fundamental Constants in Replacing the Kilogram" *IEEE Trans. Instr. Meas. 40*, 86-91

Taylor, B.N., Cohen, E.R. [1991] "How accurate are the Josephson and quantum Hall effects and QED?" *Phys. Lett. A153*, 308-312

Tsai, J.-S.; A. K. Jain; J. E. Lukens [1983], "High-Precision Test of the Universality of the Josephson Voltage-Frequency Relation", *Phys. Rev. Lett. 51*, 316-318

Verbrugh, S. [1995], "Development of a Single Electron Turnstile as a Current Standard", *Thesis, Delft, University of Technology,* 1995,

ISBN 90-9008171-2

Verbrugh, S. M.; M. L. Benhamadi; E. H. Visscher; J. E. Mooij [1995], "Optimization of island size in single electron tunneling devices: Experiment and theory", *J. Appl. Phys. 78*, 2830-2836

Verbrugh, S. M.; S. J. Tans; P. de la Court; C. J. M. Harmans; J. E. Mooij [1994], "Single Electron Turnstile as a Current Standard", *Physica B 194-195*, 1041-1042

Vion, D.; P. F. Orfila; P. Joyez; D. Esteve; M. H. Devoret [1994], "Miniature Electrical Filters for Single Electron Devices", to be published in *J. Appl. Phys.*

Visscher, E. H.; S. M. Verbrugh; J. Lindeman; P. Hadley; J. E. Mooij [1995], "Fabrication of multilayer single-electron tunneling devices", *Appl. Phys. Lett. 66*, 305

von Klitzing, K. ; G. Dorda, M. Pepper [1980], "New Method for High-Accuracy Determination of the Fine-Structure Constant Based on Quantized Hall Resistance"; *Phys. Rev. Lett. 45*, 494-497

von Klitzing, K. [1986], "The quantized Hall effect", *Rev. Mod. Phys.*, 519

Widom, A. [1979], "Quantum Electrodynamic Circuits at Ultralow Temperature", *Journ. Low Temp. Phys. 37*, 449 - 460

Widom, A.; G. Megaloudis, T. D. Clark, H. Prance, R. J. Prance [1982], "Quantum electrodynamic charge space energybands in singly connected superconducting weak links", *J. Phys. A.*: Math. Gen. 15, 3877 - 3879

Widom, A.; T. D. Clark [1982], "Quantum Electrodynamic Uncertainty Relations", *Phys. Lett. 90A*, 280

Williams, E. R.; R. N. Ghosh; J. M. Martinis [1992], "Measuring the Electron's Charge and the Fine-Structure Constant by Counting Electrons in a Capacitor", *J. Res. Nat. Inst. Stand. Technol. 97*, 299-306

Zimmerli, G.; R. L. Kautz; J. M. Martinis [1992b], "Voltage gain in the single-electron transistor", *Appl. Phys. Lett. 61*, 2616-2618

Zimmerli, G.; T. M. Eiles; R. L. Kautz; J. M. Martinis [1992a], "Noise in the Coulomb blockade electrometer", *Appl. Phys. Lett. 61*, 237-239

Zorin, A. B. [1995], "The thermocoax cable as the microwave frequency filter for single electron circuits", *Rev. Scient. Instrum. 66*, 4296-4300

Zorin, A. B.; F.-J. Ahlers; J. Niemeyer; T. Weimann; H. Wolf; V. A. Krupenin; S. V. Lothkov, "On the background Charge Noise in Metallic Single Electron Tunneling Devices", to be published in *Phys. Rev. B.*

HIGH-FREQUENCY METHODS IN ELECTROMAGNETICS

Roberto Tiberio, Stefano Maci and Alberto Toccafondi

1 Introduction

As is quite well-known, two leading theories have dominated the high-frequency scenario in the last few decades; namely the Geometrical Theory of Diffraction (GTD) [*Keller*, 1962] and the Physical Theory of Diffraction (PTD) [*Ufimtsev*, 1971].

There are certain similarities as well as significant differences in the background of these two theories. Both of them make use of solutions of local canonical problems. In most practical instances, the relevant reference canonical problems have an infinite, uniform cylindrical configuration. On one hand, in a GTD scheme the canonical problem is used to provide an estimate of a diffracted ray field contribution. On the other hand, in a PTD scheme the same canonical problems are used to find an estimate of the currents that should be introduced to correct those predicted by the Physical Optics (PO).

The GTD leads to a very attractive picture of the scattering from a complex object, in terms of rays emanating from isolated "flash points". However, as a counterpart of this simplicity, there are few drawbacks which may sometimes result in severe impairments. Searching for the relevant "flash points" in a complex environment is not at all a straightforward matter. Also, although the singularities that occur in the original GTD formulation at the Shadow Boundaries (SB) of the Geometrical Optics (GO) field have been removed by its uniform versions, such as the Uniform Geometrical Theory of Diffraction (UTD) [*Kouyoumjian* and *Pathak*, 1974, 1986] and the Uniform Asymptotic Theory [*Lee* and *Deschamps*, 1976], still there are serious difficulties in applying this ray method close to and at caustics. This problem has received a considerable attention; earlier contributions have been provided in [*Ross*, 1966] and more systematically in [*Ryan and Peters*, 1969; *Burnside* and *Peters*, 1974; *Knott* and *Senior*, 1973; *Appel-Hansen*, 1987] by the equivalent current method. However, these techniques may be applicable only at very specific aspects. Furthermore, there is an inherent limitation, at least from a conceptual point of view, due to the fact that in several cases GTD is able to predict a non-vanishing field in restricted angular regions. This latter inconvenience, may be alleviated by introducing diffracted field contributions from vertices [*Sikta et al.*, 1983; *Brinkley et al.*, 1989; *Ivrissimtzis* and *Marhefka*, 1992; *Maci et al.*, 1994; *Capolino* and *Maci*, 1996]. Nevertheless, when applicable, GTD techniques have been demonstrated to provide a very powerful method. Therefore, it would be desirable, if not imperative, for any alternative or even more general method, to be able to recover this ray field description when it is appropriate.

In principle the PTD in its more recent formulations is not affected by the same above impairments as the GTD; however, as a counterpart it may require significant numerical efforts. Important improvements have been introduced, within the framework of the original PTD pioneered by *Ufimtsev* [1971], when Incremental Length Diffraction Coefficients (ILDC) [*Mitzner*, 1974; *Shore and Yaghjian*, 1988], Equivalent Edge Currents (EEC) [*Michaeli*, 1984, 1986] and Elementary Edge Waves (EEW) [*Butorin and Ufimtsev*, 1986; *Ufimtsev*, 1989, 1991] have been formulated to asymptotically describe fringe current contributions. This typically may occur whenever an edge discontinuity introduces a relevant distortion of the PO currents. All the above formulations, which have been developed only for an edge, will be referred to as PTD techniques. The major attempt of these techniques, besides eliminating caustic singularities, is that of providing an extension for observation angles which are not on the diffraction cone.

To this end, as is also neatly illustrated by *Pathak* [1992], it should be noted that the local nature of high-frequency scattering is mainly determined by an integration process, in which the occurrence of either destructive or constructive interferences is the result of a collective phenomenon. In general, this result not only depends on the actual distribution of the currents, but is also strongly affected by the observation direction. The field, as well as the corresponding current distribution, obtained from the solution of a canonical problem which has a uniform, cylindrical configuration, is in essence a stationary phase contribution.

In our opinion the direct use of stationary phase information is the major responsible of the limitation that, in one way or another, may affect the various existing theories. The scope of obtaining a high-frequency description of the scattering phenomenon at any aspect, within a unitary, self-consistent framework, may be addressed from a somewhat different perspective.

To this end, it is shown that useful results may be achieved by extracting the desired local information directly from the field of a uniform, infinite, cylindrical configuration. It can be done by carefully examining the physical mechanism, which governs the process of building up the ray fields of these canonical problems. It is recognized that the solution of such a canonical configuration may be thought of as a superposition of an infinite uniform distribution of incremental field contributions, that are localized along a directrix of the cylinder. In other words, this radiation mechanism is governed by an element factor and by an infinite, uniform distribution factor. This concept is used to establish a Fourier transform pair relationship between the element factor and the solution of the cylindrical canonical problem. As a consequence, although the field of the canonical configuration is strongly dominated by its infinite, uniform distribution factor, this procedure, which will be referred to as a generalized localization process, allows to extract the pertinent element factor. The associated incremental field contribution explicitly satisfies reciprocity and its local boundary conditions (b.c.).

It is shown that this high-frequency method, which is referred to as an Incremental Theory of Diffraction (ITD) [*Tiberio and Maci*, 1994; *Tiberio et. al.*, 1995, *Maci et al.*, 1996a] may provide a quite general, self-consistent procedure to describe the various scattering phenomena at high-frequency within a unitary framework, which naturally includes the UTD ray field representation when it is applicable.

The presentation is organized as follows. First, in Sect. 2 a conceptual background of high-frequency methods is provided. In particular, the foundations of the GTD and its uniform extensions are quickly addressed, to recall its successful applicability to a variety of practical and complex engineering problems. Also, a brief mention of the recent developments of the PTD is given in this same section. The ITD formulation for wedge shaped configuration is summarized in Sect. 3. Its extension and generalization to treat local cylindrical configurations with a curved cross-section is presented in Sect. 4. The PO, as involved in the total field representation of the incremental theories, is addressed in Sect. 5, where its relationship with the Aperture field Integration (AI) is discussed.

2 Background

The electromagnetic radiation and scattering in general may be described by the radiation integral of the currents induced over the scatterer. At lower frequencies, one may determine these currents by solving an integral equation that represents the boundary conditions (method of moments, conjugate gradient methods, etc.). Alternatively one may solve directly the differential equations governing the total field behavior (finite element or finite difference methods).

At high-frequency, namely when the object is large in terms of a wavelength, all these numerical techniques become poorly convergent and inefficient; therefore, the use of high-frequency techniques becomes necessary. In those circumstances, electromagnetic scattering and diffraction exhibit a "highly localized" behavior. In order to clarify the meaning of this concepts, let us consider the scattered field in terms of the radiation integral

$$\vec{E}^{s}(\vec{r}',\vec{r}) = \iint_{S} \vec{B}(\vec{r}_s)\, e^{-jkf(\vec{r}_s)}\, dS \tag{1}$$

in which S is the surface of the object, $\vec{r}'$ denotes the position of a perfectly focalized source, $\vec{r}$ and $\vec{r}_s$ the point of observation and integration, respectively. If there are no multiple reflection mechanisms, the high-frequency principle of localization defines the phase function in (1) as the distance between the observation and the source point constrained to the surface; i.e.

$$f(\vec{r}_s) = \left|\vec{r}'_s - \vec{r}_s\right| + \left|\vec{r} - \vec{r}_s\right| \tag{2}$$

This assumption may be used to represent the scattered field as a superposition of elementary spherical waves emanating from each point $\vec{r}_s$ of the surface; each contribution is excited by a phase term which is dictated by the illuminating field. The complex amplitude $\vec{B}$ of these elementary waves is *a priori* unknown. At high-frequency, these contributions produce destructive interferences except for those arising from the neighborhood of stationary phase points $\vec{r}_{sm}$ on the surface and from the boundary l of the integration domain. This description leads to the following asymptotic representation of the scattered field

$$\vec{E}^{s} \sim \sum_{m} \frac{\vec{B}(\vec{r}_{sm})}{\Delta_1(\vec{r}_{sm})}\, e^{-jk\, f(\vec{r}_{sm})} + \int_{l} \frac{\vec{B}(\vec{r}_l)}{\delta(\vec{r}_l)}\, e^{-jk\, f(\vec{r}_l)}\, dl \tag{3}$$

The first summation represents contributions coming from "flash points" of the first kind that correspond to the optic specular points. Each contribution is a reflected ray and the summation represents the well-known GO field.

The boundary contribution from the line (l) integral is the superposition of Incremental Boundary Waves (IBWs) coming from Shadow Boundary Lines (SBLs) and geometrical or electrical edge discontinuities on the surface. The IBWs produce destructive interferences except again for the stationary phase points $\vec{r}_{ln}$ and the end-point contributions $\vec{r}_{vh}$. Eventually, the edge wave field may be represented as a superposition of diffracted rays

$$\int_l \frac{\vec{B}(\vec{r}_l)}{\delta(\vec{r}_l)} e^{-jk f(\vec{r}_l)} dl = \sum_{n,h} \left[\frac{\vec{B}(\vec{r}_{ln})}{\Delta_2(\vec{r}_{ln})} e^{-jkf(\vec{r}_{ln})} + \frac{\vec{B}(\vec{r}_{vh})}{\Delta_3(\vec{r}_{vh})} e^{-jkf(\vec{r}_{vh})} \right] \quad (4)$$

originating from "flash points" of the second ($\vec{r}_{ln}$) and third ($\vec{r}_{vh}$) kind. In certain circumstances the asymptotic representations (1) and (2) are not definable. It happens when stationary phase conditions realize on continuous portions of the actual surface (line). The corresponding locus of observation points is a caustic of reflected (diffracted) rays. Near caustics the ray description of the scattered field looses generality and in most instances the asymptotic field needs to be evaluated by direct integration of incremental contributions.

2.1 Geometrical Theory of Diffraction

The GTD has been developed within the framework of a field representation which consists of the sum of the GO field plus diffracted ray field contributions such as those in (4). Accordingly, the various ray field contributions are estimated from the solution of canonical configurations, which locally fit that of the actual object in a small neighborhood of the stationary points. To determine these scattering centers and their relevant diffracted ray-field contributions, Keller postulated a generalization of the Fermat's principle. This generalization states that an optical (stationary) ray path may be established through points located on the scattering object, where the GO field is discontinuous. Typically, such points belong to shadow boundaries, edges and tips.

Thus, high-frequency solutions of local canonical problems are basic constituents for the applicability of this ray method. Formulations of these solutions, that are uniformly valid through the GO shadow boundaries, provided a major breakthrough to achieve a definitive impact on treating practical engineering configurations. Excellent overviews of the major achievements and most recent developments in modern GTD may be found in [*Pathak*, 1992] and in [*Bouche et al.*, 1993]. On these bases, general purpose codes have been developed that are successfully being used worldwide.

However, still there are certain zones where even the uniform extensions of GTD fail because the ray field representation is intrinsically invalid, such as boundary layers and caustics. To this end, it should be noted that the reference solution of a local canonical problem, which in most instances has a uniform cylindrical configuration, essentially describes a stationary phase field contribution. As a consequence, the GTD estimate of such ray field is applicable only at those incidence and observation aspects, where an isolated stationary phase condition is well-established.

2.2 Physical Theory of Diffraction

At a variance, the PTD is based on a representation of the integrand in (1) as the sum of a contribution $\vec{B}^{po}$, obtained from a PO approximation of the local currents on the lit region of the object, plus a contribution $\vec{B}^{nu}$ from non uniform currents, that are introduced in addition to those predicted by PO. Since the exact currents are unknown, one needs to seek for an estimate of the non uniform currents such that $\vec{B}^{nu} \sim \vec{B} - \vec{B}^{po}$

The basic assumption of PTD consists of assuming that the non uniform currents are locally determined by the diffracted field from the same local canonical configurations, that are used in GTD to directly estimate the ray field contributions. For certain canonical wedge configurations, the PTD non uniform currents that are distributed on a face of the local canonical wedge, can be integrated in a closed form along one of the coordinates which may be used to describe the local face. This procedure leads to define EEW contributions that after being integrated along the edge of the canonical wedge, reconstruct the original exact solution. To this end, it is assumed that the EEW is the field radiated by an elementary strip of currents along the local ray path of the GTD diffracted field. It is rather apparent that one may choose different orientations for the strips. However, in doing that it is found that a cone of non physical singularities occurs in the expression for the incremental length field contribution. This above specific choice of the strip orientation, besides removing this latter severe mathematical inconvenience, has its own physical motivation as clearly explained in [*Ufimtsev*, 1991]. There one reads *"The elementary strips of the wedge faces are to be oriented in such a way that a scattered field on such strips should contain the edge waves coming only from the "self" element of the wedge edge. It appears that only the elementary strips oriented along the diffraction rays [...] have such property. Indeed, for any other orientations of the strips, the field on their surface will be produced by the edge waves (diffraction rays) coming from the "strange" elements of the wedge edge. It is obvious that in the case of a curved (not rectilinear) edge, such strange elements of the wedge edge are fictitious. Consequently, any other orientation of the elementary strips [...] is absurd from a physical point of view."*

It should be noted that, although these currents are assumed to be associated to an elementary local contribution, they are essentially determined by the collective, stationary phase contribution of the local, infinite canonical problem. When the EEW contributions are distributed and integrated along the actual edge, they provide a fringe field augmentation of the PO field.

According to the original formulation, non uniform currents should be defined on the local surface of the scatterer. However, recently it has been suggested [*Michaeli*, 1995] that the same idea may be extended to aperture fields to provide a fringe augmentation of the Kirchhoff-Kottler aperture integration.

2.3 A recent development

A different point of view is adopted by the ITD to devise an alternative, self-consistent method for describing high-frequency scattering phenomena. To this end, as mentioned earlier, Fourier Transform (F.T.) pair relationships are applied to the solution for the diffracted field of a canonical problem, which has an infinite, uniform cylindrical configuration. This localization process, which is applicable to any canonical, cylindrical configuration with arbitrary cross-section, is used to define local incremental diffracted field contributions. Next, they are adiabatically distributed along the actual edges and boundary lines of the GO field, to provide an estimate of the second term in (3). Then, an estimate $\vec{B}^{go}$ of $\vec{B}$ in (1) is used such as that from the PO approximation. This integration process allows to recover the GO ray field in (3) by its stationary point contributions. However, inaccurate end-point contributions are also introduced by this procedure. These latter are estimated from a local canonical problem consisting of a semi-infinite distribution of $\vec{B}^{go}$, and the same localization process is applied to its field. When such contributions are subtracted from that of the surface integration of $\vec{B}^{go}$, a field is obtained which asymptotically contains only the GO field information. Thus, this field representation directly provides the GTD ray field contributions, when the relevant stationary phase approximations are applicable. It should also be noted that the same procedure may be used to find incremental fringe fields to directly augment the PO field. However, both the formulation and the conceptual background of these latter fringe contributions are different from those of the EEW.

3 ITD for wedge shaped configurations

Let us consider an arbitrarily shaped, opaque object which is large in terms of a wavelength and let us focus our attention on the SBL which occurs at an edge discontinuity.

At each point Q' along the edge *l*, a locally tangent, wedge canonical problem is defined which has an infinite, uniform, cylindrical configuration (Fig. 1). Let us introduce a cylindrical coordinate system (ρ, ϕ, z) with the z axis at the edge of the wedge, and a spherical (r, β, ϕ) coordinate system, with its origin at Q'. Also, we denote by (β', ϕ') the local direction of incidence at Q', and by $n\pi$ the exterior wedge angle.

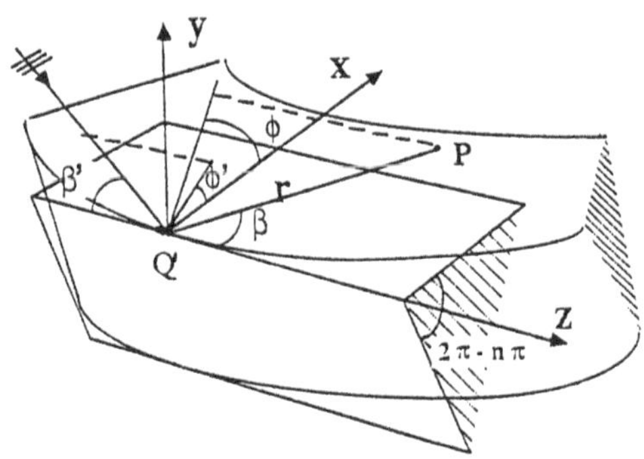

Fig. 1. Geometry at a locally tangent wedge.

The infinite canonical structure at Q' exhibits an infinite directivity with respect to the β angular coordinate, which may allow a direct estimate of the field only in those directions belonging to the diffraction cone $\beta = \beta'$.

However, as mentioned earlier, an incremental field contribution is localized at any point along the edge, and is determined by extracting it from the locally tangent, uniform cylindrical canonical problem, via the localization process which is specified herein after. To this end, for the sake of simplicity in the discussion but without loss of generality, let us consider the scalar problem.

The solution $\psi^d(\rho, \phi, z, \beta')$ for the diffracted field of the canonical problem at any point P has the plane wave spectral integral representation

$$\psi^d(\rho,\phi,z,\beta') = \frac{1}{2\pi j}\int_{C_{\alpha'}} G(\alpha',\phi')\,\psi^i\, e^{-jk[\rho\cos(\alpha' \mp \phi)\sin\beta' + z\cos\beta']}d\alpha' \tag{5}$$

in which $\mp$ signs apply for $\phi \gtrless \pi$. Furthermore, ψ^i is the incident field at z = 0, and G is the well-known spectrum function which is expressed by the rather standard combination of four cotangents. The contour of integration C ≡ (-i∞, π + j∞), in order to isolate the diffracted field, is indented in such a way that the GO pole is not captured, when it is deformed onto a steepest descent path. Due to the cylindrical uniformity of this wedge canonical configuration, the above solution (5) may be represented as the superposition of an infinite distribution of incremental contributions

$F^d(Q')$, uniformly distributed along the edge. Thus, a F.T. pair relationship is established between $F^d(Q')$ and ψ^d. Accordingly,

$$F^d(Q') = \frac{k}{4\pi^2 j} \int_{C_{\theta'}} \int_{C_{\alpha'}} G(\alpha', \phi') \psi^i(Q') \; e^{-jkr(\sin\theta' \sin\beta \cos(\alpha' \mp \phi) + \cos\theta' \cos\beta)} \sin\theta' \, d\alpha' d\theta' \tag{6}$$

is the desired plane wave, double spectral integral representation for the localized, incremental field contribution at any point P(r, β, ϕ) from any point Q' along the edge l, where a zero phase reference is assumed. These incremental fields are used as element factors of a linear radiating structure with a distribution factor, which is determined by both the actual geometry of the edge and the local illumination.

The plane wave spectral integral representation (6) of the incremental field contribution, is asymptotically evaluated for kr large to yield closed form high-frequency expressions. The explicit formulation will be given later on for the electromagnetic vector problem. It will be seen that the incremental contribution to the diffracted field can be calculated at high-frequency by rather familiar functions, typically involved in the UTD.

In order to obtain a convenient total field representation, let us now refer to surface incremental field contributions that may be estimated from a physical (PO) equivalent approximation. Those contributions F^{go}_{s}, are distributed and integrated on their relevant actual lit surface. This procedure provides a field which recovers the corresponding GO field by its stationary phase contributions, when they exist in the integration process. In addition, the integration process asymptotically introduces end-point contributions from the SBLs. These latter are inaccurate because they do not contain the correct local information at the edge of the SBL. Thus, they should be subtracted and replaced by proper incremental diffracted field contributions, such as those introduced above for the edge of a local wedge.

The improper incremental GO end-point contributions may be asymptotically estimated as follows. Let us consider a point Q' at the SBL l of the lit surface S_l of an object. A pertinent local canonical problem is that of an infinite plane, with an infinite straight SBL locally tangent to the actual SBL at Q'. It is useful to introduce a coordinate system at the SBL, which is congruent with that chosen for the edge problem. Next, the surface incremental GO field is integrated all over the semi-infinite extent $x \geq 0$. This yields a field contribution which arises from a uniform cylindrical configuration along the z axis, and thus exhibits an infinite directivity with respect to the β coordinate for $\beta=\beta'$. Such a field is well-suited for applying to it the same localization process as that introduced for the edge of a wedge. This leads to define localized incremental, end-point GO contributions F^{go}_{e} (Q').

Again, these incremental fields are adiabatically distributed and integrated along the actual SBL. In subtracting this latter contribution denoted by ψ^{go}_{l} from that of the surface integration denoted by ψ^{go}_{s} a generalized GO field is obtained as

$$\psi^{go}_{g} = \psi^{go}_{s} - \psi^{go}_{l} \tag{7}$$

The total field ψ_t is then obtained by introducing the contribution ψ^{d}_{l} from F^{d} along the same SBL, so that

$$\psi_t = \psi^{go}_{g} + \psi^{d}_{l} \tag{8}$$

The above scalar formulation can be extended without any major difficulty to treat the electromagnetic vector problem. For the sake of completeness, explicit dyadic high-frequency expression for the incremental field contributions are given herein after. The incremental diffracted field contributions are calculated as

$$\vec{f}^{d}(Q') = \overline{\overline{S}}^{d}(Q') \cdot \vec{E}^{i}(Q') \frac{e^{-jkr}}{2\pi r} \tag{9}$$

in which $\vec{E}^{i}$ is the incident electric field at Q', and

$$\overline{\overline{S}}^{d} = D^{s}\hat{\beta}\hat{\beta}' + D^{h}\hat{\phi}\hat{\phi}' \tag{10}$$

where

$$D^{s}_{h} = -\sum_{i=1}^{2} \in_{s,h} \frac{1}{2n} \cot\left[\frac{\pi + \delta_j\phi + \delta_i\phi'}{2n}\right] \mathfrak{F}\left[2K\cos^2\left(\Phi_{ij}/2\right)\right] \tag{11}$$

with $\delta_l = (-1)^l$, $\in_s = -1$ and $\in_h = \delta_i\delta_j$. Furthermore, in (11),

$$K = k\frac{r\sin^2\beta \,.\, r'\sin^2\beta'}{r\sin^2\beta + r'\sin^2\beta'} \tag{12}$$

$$\Phi_{ij} = \pi + \delta_j\phi + \delta_i\phi' - 2n\pi N_{ij} \tag{13}$$

and $\mathfrak{F}(\,.\,)$ is the UTD transition function. When the PO approximation is used, the incremental end-point contributions are calculated as

$$\vec{f}^{po}_{e}(Q') = \overline{\overline{S}}^{po}_{e}(Q') \cdot \vec{E}^{i}(Q') \frac{e^{-jkr}}{2\pi r} \tag{14}$$

in which

$$\vec{\vec{S}}^{po}_{e} = u(\pi - \phi')(S^{po}_{\beta\beta'}\hat{\beta}\hat{\beta}' + S^{po}_{\beta\phi'}\hat{\beta}\hat{\phi}' + S^{po}_{\phi\phi'}\hat{\phi}\hat{\phi}') \tag{15}$$

where

$$S^{po}_{\substack{\beta\beta'\\ \phi\phi'}} = \sum_{i=1}^{2} \varepsilon_{\substack{\beta\beta'\\ \phi\phi'}} \frac{1}{2} \cot\left[\frac{\pi - (\phi + \delta_i \phi')}{2}\right] \mathcal{F}\left[2K\cos^2\left(\frac{\phi + \delta_i \phi'}{2}\right)\right] \tag{16}$$

with $\varepsilon_{\beta\beta'} = 1$ and $\varepsilon_{\phi\phi'} = \delta_i$ in the soft and hard cases, respectively and

$$S^{po}_{\beta\phi'} = -\frac{\cos\beta\,(\cos\phi + \cos\phi')}{\sin\phi} S^{po}_{\phi\phi'} \tag{17}$$

The above high-frequency expressions involve the same transition functions; thus, it is seen that both (9) and (14) are discontinuous at the same shadow boundary planes. However, such discontinuities precisely compensate each other. It is also worth pointing out that in the extreme far zone ($kr \to \infty$) the transition functions become unity, and both $f^d(Q')$ and $f^{go}_e(Q')$ exhibit singularities, that precisely cancel out each other.

The representation of the high-frequency field in terms of a generalized GO field plus incremental diffracted fields is continuous and well-behaved under any conditions, including caustics and SBs of the corresponding ray field representation. It is also rather apparent that the field representation introduced here naturally leads to the GTD/UTD ray field representation, when it is applicable. It is worth emphasizing that the formulation introduced in this paper is also well-suited to fit a fringe description of that field contribution $\psi^f_l = \psi^d_l - \psi^{go}_l$ which is intended to correct ψ^{go}_s. It is basically just a matter of terminology. There is, however, a substantial difference in the procedure for evaluating the incremental contributions, as well as a conceptual difference in the basic assumption behind it. Also, the final expressions for the incremental field contributions are different from those predicted by the EEW developed from PTD. Therefore, the method presented here leads to introducing different basis functions for describing high-frequency scattering phenomena.

4 ITD formulation for circular cylinders

In this section a new ITD treatment of circular cylinder-shaped structures is presented. A quite general formulation is developed which is based on a suitable interpretation of the exact solution for the scattering in the near zone from a canonical circular cylinder, when it is illuminated by a spherical source. It is shown that its spectral integral representation may also be represented as an integral convolution along the axis of the cylinder. This leads to directly define the local incremental field contribution.

Explicit expressions for incremental scattering coefficients are obtained that are applicable to moderately sized local cylinders. For the sake of simplicity, the scalar cases for both hard and soft b.c. are treated here. In this same scalar case, the formulation is easily extended to impedance b.c. Its extension to the more general vector problem of electromagnetic wave scattering from perfectly conducting circular cylinders is a relatively simple matter. It is suggested that the same general formulation is applicable to treat the case of local circular cylinders that are large in terms of a wavelength; however, it requires a more involved asymptotic analysis.

Let us consider an object with circular cross-section along a curved line l, and assume that the local radius of curvature along l is large in terms of a wavelength. At high-frequency, the scattering is a local phenomenon which in the present example may be described by locally approximating the actual object by an infinite circular cylinder with its z axis locally tangent to l at any point Q_l. The local radius of the cylinder is denoted by a. Both a cylindrical (ρ, ϕ, z) and a spherical (r, β, ϕ) coordinate systems are defined, with their origin at Q_l.

Let us analyze the scalar case of a canonical circular cylinder with either soft (s) or hard (h) b.c., which is illuminated by a point source at $P' \equiv (r', \beta', \phi') = (\rho', \phi', z')$. The exact spectral integral representation [*Felsen*, 1994] of the scalar scattered field at any point $P \equiv (r, \beta, \phi) = (\rho, \phi, z)$ is

$$G^{s,h}(\vec{r},\vec{r}') = -\frac{j}{4}\sum_{n=-\infty}^{\infty} e^{-jn(\phi-\phi')} G_n^{s,h}(\vec{r},\vec{r}') \tag{18}$$

where

$$G_n^{s,h}(\vec{r},\vec{r}') = \frac{1}{2\pi}\int_{-\infty}^{\infty} \overline{G}_n^{s,h}(\rho,\rho';k''_\rho) e^{-jk''_z(z-z')} dk''_z \tag{19}$$

in which

$$\overline{G}_n^{s,h}(\rho,\rho';k''_\rho) = C_n^{s,h}(k''_\rho a) H_n^{(2)}(k''_\rho \rho) H_n^{(2)}(k''_\rho \rho') \tag{20}$$

$$k''_\rho = \sqrt{k^2 - (k''_\rho)^2} \tag{21}$$

$$C_n^s\left(k''_\rho a\right)=\frac{J_n\left(k''_\rho a\right)}{H_n^{(2)}\left(k''_\rho a\right)};\; C_n^h\left(k''_\rho a\right)=\frac{J'_n\left(k''_\rho a\right)}{H'^{(2)}_n\left(k''_\rho a\right)}, \tag{22}$$

J_n and $H_n^{(2)}$ denote cylindrical Bessel functions of the first kind and Hankel functions, respectively; furthermore, in (22) primed functions denote derivatives with respect to their argument. Again an $\exp(j\omega t)$ time dependence is assumed and suppressed.

Our purpose is to find a local incremental scattered field contribution $\psi(Q_l)$ to be distributed and integrated along to give the total scattered field $\Psi(P)$ from the actual object.

$$\Psi(P)=\int_l \psi(Q_l)dl \tag{23}$$

According to the ITD, the incremental field $\psi(Q_l)$ is deduced from the above cylindrical, local canonical problem at Q_l provided that its solution may be represented as

$$G(\vec{r},\vec{r}')=\int_{-\infty}^{\infty} \psi(z'')dz'' \tag{24}$$

where for the sake of simplicity superscripts *s,h* are omitted. A representation as that indicated in (24) was exploited by *Bateman* [1955] and extended by *Rubinowicz* [1965] for the spectral Green's function of the wedge. However, both their purpose and their procedure are different from those discussed hereinafter.

To this end, it is useful to employ in (19) the identity

$$2C_n^{s,h}\left(k''_\rho a\right)=1+e^{j2\Theta_n^{s,h}\left(k''_\rho a\right)} \tag{25}$$

in which

$$\Theta_n^s(K)=\tan^{-1}\left[\frac{Y_n(K)}{J_n(K)}\right];\Theta_n^h(K)=\tan^{-1}\left[\frac{Y'_n(K)}{J'_n(K)}\right] \tag{26}$$

where $Y_n(K)$ and $Y'_n(K)$ denote cylindrical Bessel functions of the second kind and their derivatives with respect to the argument, respectively. As a consequence, $\bar{G}_n$ may be rewritten as

$$\overline{G}_n\left(\rho,\rho' ;k''_\rho\right)=\frac{1}{2}\left[U_n^{(1)}\left(k''_\rho\rho\right)U_n^{(1)}\left(k''_\rho\rho'\right)+U_n^{(2)}\left(k''_\rho\rho\right)U_n^{(2)}\left(k''_\rho\rho'\right)\right] \quad (27)$$

where

$$U_n^{(1)}(k_t\tau)=H_n^{(2)}(k_t\tau)\;;U_n^{(2)}(k_t\tau)=H_n^{(2)}(k_t\tau)e^{j\Theta_n(k_ta)} \quad (28)$$

After introducing (26) into (18), it is found that

$$G_n(\vec{r},\vec{r}\,')=G_n^{(1)}(\vec{r},\vec{r}\,')+G_n^{(2)}(\vec{r},\vec{r}\,') \quad (29)$$

in which, by F.T. analysis,

$$G_n^{(i)}(\vec{r},\vec{r}\,')=\frac{1}{2\pi}\int_{-\infty}^{\infty}U_n^{(i)}\left(k''_\rho\rho\right)U_n^{(i)}\left(k''_\rho\rho'\right)e^{-jk''_z(z-z')}\,dk''_z$$

$$\int_{-\infty}^{\infty}u_n^{(i)}(z''-z)\,u_n^{(i)}(z'-z'')\,dz'' \quad (i=1,2) \quad (30)$$

where $u_n^{(i)}(z''-z)$ and $u_n^{(i)}(z'-z'')$ are the inverse F.T. of $U_n^{(i)}(k_\rho\rho)$ and $U_n^{(i)}(k'_\rho\,\rho')$, respectively. Using (29) and (30) in (18) leads to the desired representation indicated in (24). Thus, it is rather apparent that the incremental scattered field contribution at Q_l $(z''=0)$ is

$$\psi(Q_l)=\frac{j}{4}\sum_{n=-\infty}^{\infty}e^{-jn(\phi-\phi')}\left[\psi_n^{(1)}(Q_l)+\psi_n^{(2)}(Q_l)\right] \quad (31)$$

where

$$\psi_n^{(i)}(Q_l)=\frac{1}{(2\pi)^2}\int_{-\infty}^{\infty}\int_{-\infty}^{\infty}U_n^{(i)}(k_\rho\rho)U_n^{(i)}(k'_\rho\rho')e^{-j(k_zz-k'_zz')}dk_z\,dk'_z \quad (32)$$

It is worth pointing out that expression (31) explicitly satisfies reciprocity.

In order to derive more tractable expressions for the incremental scattered field, let us first consider expression (32) with i=1. It consists of the product of two typical integral expressions, such as

$$u_n^{(1)}(\zeta)=\frac{1}{2\pi}\int_{-\infty}^{\infty}H_n^{(2)}(k_t\tau)e^{jk_\zeta\zeta}dk_\zeta\;,\;k_t=\sqrt{k^2-(k_\zeta)^2} \quad (33)$$

A closed form expression for (33) is then obtained by employing the following symptotic approximation:

$$2\,h_n^{(2)}(k\eta) \sim \int_{-\infty}^{\infty} H_n^{(2)}(k_t\tau)\,e^{jk_\zeta\zeta}\,dk_\zeta \tag{34}$$

in which $\eta^2 = \tau^2 + \zeta^2$, and $h_n^{(2)}(k\eta)$ are spherical Hankel functions.

Let us next consider expression (32) with i=2. It involves the typical integral

$$u_n^{(2)}(\zeta) = \frac{1}{2\pi}\int_{-\infty}^{\infty} H_n^{(2)}(k_t\tau)\,e^{j\theta_n(k_t a)}\,e^{jk_\zeta\zeta}\,dk_\zeta\,,\quad k_t = \sqrt{k^2-(k_\zeta)^2} \tag{35}$$

Provided that $\exp[j\Theta_n(k_t a)]$ is a slowly varying function, a stationary phase approximation of (35) yields

$$u_n^{(2)}(\zeta) \sim e^{j\theta_n(k\sin\theta a)}\,u_n^{(1)}(\zeta) \tag{36}$$

so that, by using (34), an approximate closed form expression is obtained for (35). Finally, the closed forms for (33) and (35) are used in (31) so that, after introducing the useful notation

$$kh_n^{(2)}(kr) = j\,\frac{e^{-jkr}}{r}\,\overline{h}_n^{(2)}(kr) \tag{37}$$

with $\overline{h}_n^{(2)}(kr) = \Sigma_{m=0}^{n}(n+1/2,m)j^{n-m}(2kr)^{-m}$, the incremental scattered field is conveniently expressed in terms of a scattering coefficient as

$$\psi(Q_l) \sim \frac{e^{-jkr'}}{r'}\,S(\vec{r},\vec{r}\,';a)\,\frac{e^{-jkr}}{r} \tag{38}$$

At high-frequency, neglecting terms of higher order than $(kr)^{-1}(kr')^{-1}$ yields an expression of the scattering coefficient that is independent of kr and kr'.

It should also be emphasized that in order to obtain the above closed form expression (38), the basic approximation has been introduced which led from (35) to (36). As a consequence, it is found that expression (38) is applicable to small and moderately sized circular cross-sections. When the radius becomes large in terms of a wavelength, the present solution fails. In this case, it is more appropriate to refer to the alternative exact expression of the Green's function, in which the summation in (18) is replaced by its corresponding integration [*Felsen*, 1994]. It can be shown that a quite similar formulation for the incremental scattered field may be obtained; however, reducing it to practical expressions requires a rather involved asymptotic analysis.

5 Equivalence between physical optics and aperture integration

So far, it is rather apparent that in one way or the other, PO is a constituive part of incremental theories. For certain configurations, the Kirchhoff-Kottler AI technique is used as an effective alternative to calculate useful first order approximations of the radiated field. Even though AI and PO approximations are very simple and well-established high-frequency techniques, the relationship between them was not yet fully exploited in the literature. The equivalence between these two techniques was previously found only in the far zone for parallel plate and rectangular open-ended waveguides (OEWs) [*P. Ya. Ufimtsev, unpublished paper,* 1986]. Just very recently [*Maci et al.*, 1996b], this equivalence has rigorously been established for OEWs of arbitrary cross-section at any observation point. One should also note an interesting communication [*Oodo and Ando*, 1995] where a numerical equivalence between PO and AI far fields was established for large reflector antennas.

In this section, the formulation of this equivalence is summarized, which is relevant to the high-frequency description of the radiation of eigen-mode field configurations from an OEW. As is well-known, the above mentioned basic approximations are often used for their simplicity. They provide the radiated field by resorting to the free-space radiation integral of the equivalent currents relevant to unperturbed mode distributions; in either AI or PO techniques, the current distributions are defined either on the aperture or on the walls, respectively. Despite of their simplicity, these techniques are found accurate for most practical applications, also for moderate waveguide dimensions, except may be for wide-angle observation and below the aperture plane where fringe current and multiple diffraction contributions cannot be neglected.

Although a rigorously established equivalence between AI and PO techniques is interesting *per se*, in practical applications it also provides the freedom of choosing that technique which allows more efficient calculations. In most instances, AI is a more desirable formulation which implies a finite domain integration. However, owing to the semi-infinite extent of its integration domain, sometimes the PO integration can be easily performed in a closed form along the generatrix of the cylinder; thus, it allows AI to be exactly reduced to a line integration along the edge.

In order to rigorously formulate the above equivalence, let us consider a cylindrical OEW with arbitrary cross-section and perfectly conducting walls. The waveguide is excited by an arbitrary mode traveling toward the positive z direction; its electric and magnetic fields are denoted by $\vec{E}_{mod}$ and $\vec{H}_{mod}$, respectively. Let us assume the aperture plane at $z = 0$, and denote by V_{in} and V_{ex} the volume internal and external to the OEW, respectively, as depicted in Fig. 2.

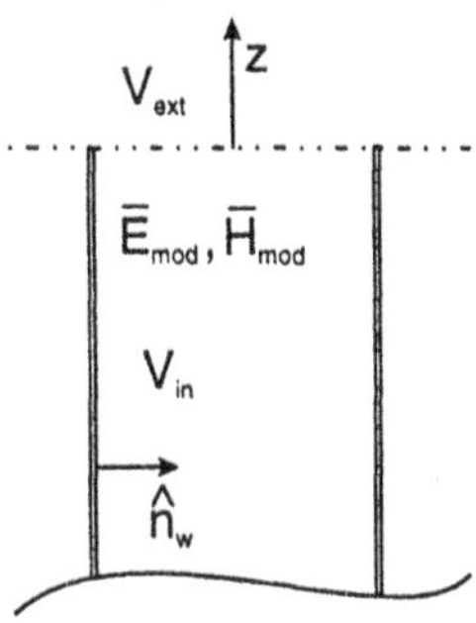

Fig. 2. Geometry at an open-ended waveguide

The PO currents J_{po} on the walls are defined as

$$\vec{J}_{po} = \hat{n}_w \times \vec{H}_{mod} \Big|_{z<0} \tag{39}$$

in which $\hat{n}_w$ is the normal to the wall directed toward V_{in}. These currents radiate at any point P in free space the PO field ($\vec{E}_{po}$, $\vec{H}_{po}$).

The Kirchhoff-Kottler approximation assumes equivalent currents

$$\begin{aligned} \vec{J}_{ka} &= \hat{z} \times \vec{H}_{mod} \\ \vec{M}_{ka} &= \vec{E}_{mod} \times \hat{z} \end{aligned} \tag{40}$$

on the aperture, that radiate a field ($\vec{E}_{ka}$, $\vec{H}_{ka}$) at any point P in the free space. The following theorem has been demostrated [*Maci et al.*, 1996b]

$$\left(\vec{E}_{po}, \vec{H}_{po}\right) \equiv \left(\vec{E}_{ka}, \vec{H}_{ka}\right) \qquad P \in V_{ex} \tag{41}$$

$$\left(\vec{E}_{po}, \vec{H}_{po}\right) \equiv \left(\vec{E}_{ka}, \vec{H}_{ka}\right) + \left(\vec{E}_{mod}, \vec{H}_{mod}\right) \qquad P \in V_{ex} \tag{42}$$

i.e., the AI and the PO fields are exactly the same outside the open ended waveguide; inside the same waveguide the sum of the AI field plus the modal field is equal to the PO field.

Note that, outside the OEW (in V_{ex}), the PO field is a radiated field; to evidence this, it can be denoted by ($\vec{E}_{po}^{\,rad}$, $\vec{H}_{po}^{\,rad}$). Inside the OEW (in V_{in}) the PO field is the sum

$$\begin{aligned} \vec{E}_{po} &= \vec{E}_{mod} + \vec{E}_{po}^{\,ref} \\ \vec{H}_{po} &= \vec{H}_{mod} + \vec{H}_{po}^{\,ref} \end{aligned} \tag{43}$$

where ($\vec{E}_{mod}$, $\vec{H}_{mod}$) is the incident field and ($\vec{E}_{po}^{\,ref}$, $\vec{H}_{po}^{\,ref}$) is the field reflected back from the end. This latter contains all the eigen-modes. Then, we can interpret the theorem (41), (42) as follows

$$\left(\vec{E}_{ka}, \vec{H}_{ka}\right) \equiv \begin{cases} \left(\vec{E}_{po}^{rad}, \vec{H}_{po}^{rad}\right) & \text{for} \quad P \in V_{ex} \\ \left(\vec{E}_{po}^{ref}, \vec{H}_{po}^{ref}\right) & \text{for} \quad P \in V_{in} \end{cases} \tag{44}$$

This means that the AI field in the Kirchhoff-Kottler approach exactly equals the PO radiated and reflected fields.

It is worth noting that the assumption of perfectly conducting walls, is not at all necessary. This theorem is applicable to any arbitrary either boundary conditions or configurations of its walls, provided that in this special case the incident dominant waveguide mode of the original, canonical infinite waveguide may be assumed entirely confined within the waveguide itself. Also, it is applicable for any arbitrary modal field configuration which is compatible with the canonical infinite waveguide.

The above theorem may be extended to deal with dominant modes in waveguides smoothly expanded into horns, provided that in this special case it is assumed that the incident waveguide mode adiabatically transforms into the lowest order eigen-mode of the horn without reflection and excitation of other modes. The fields $(\vec{E}_{po}^{rad}, \vec{H}_{po}^{rad})$ and $(\vec{E}_{po}^{ref}, \vec{H}_{po}^{ref})$ are found by the Kirchhoff-Kottler aperture integration.

Finally, it is suggested that applying this theorem allows to more easily conceive an asymptotic reduction of AI into line integration, based on a rigorous high-frequency approximation of the PO integration along the semi-infinite generatrix of the OEW structure.

An important consequence of the above theorem within the framework of incremental theories is that of allowing to complement either AI or PO [*Michaeli*, 1995; *Capolino et al.*, 1995] by the same fringe contributions. We just note that the fringe field that can be deduced from ITD, as shown in Sect. 3, does not depend on any choice of the surface where the equivalence principle is applied. Thus, it is rather apparent that the ITD formulation is congruent with the above theorem.

6 Conclusions

A brief overview has been presented of well-established high-frequency methods form treating antenna and scattering problems. The attention has been focused on incremental techniques, that may provide useful extensions of either ray field or integration techniques. Recent developments within this framework have been discussed in some details. A relationship between PO and AI has been addressed, which may be useful to provide an effective representation of the total field, after conveniently introducing either diffracted or fringe incremental contributions.

References

Appel-Hansen, J. [1987], "Equivalent edge currents", 1987 SMBO International Microwave Symposium, Rio de Janeiro, Brasil, 1, 379-384.

Bateman, H. [1955], *The Mathematical Analysis of Electrical and Optical Wave Motion on the Basis of Maxwell's Equation.* Dover, 90-94.

Bouche, D. P., F. A. Molinet and R. Mittra [1993], "Asymptotic and hybrid techniques for electromagnetic scattering", *Proc. IEEE*, 81, 1658-1684.

Brinkley, T. J., L. P. Ivrissimtzis and R. J. Marhefka [1989], "Uniform theory of diffraction analysis of electromagnetic scattering from corners and vertices," Technical Report 312506-1, *The Ohio State University ElectroScience Laboratory, Department of Electrical Engineering.* Prepared under Contract No. Z90115 for McDonnell Douglas Corp.

Burnside, W. D. and L. Peters, Jr. [1974], "Edge diffracted caustic fields," *IEEE Trans. Antennas Propagat.*, AP-22, 620-623.

Butorin, D. I. and P. Ya. Ufimtsev [1986], "Explicit expressions for an acoustic edge wave scattered by an infinitesimal edge element," *Sov. Phys. Acoust.*, 32, 283-287.

Capolino, F., S. Maci, F. Mioc, R. Tiberio [1995], "ITD approach for predicting near field radiation by a circular horn" *IEEE AP-S Symposium*, Newport Beach, California, 1, 548-551.

Capolino, F., and S. Maci [1996], "Uniform high-frequency description of singly, doubly, and vertex diffracted rays for a plane angular sector" to be published on *Jour. on Electromag. Wave and Applic.*

Felsen, L. and N. Marcuwitz [1994] "Radiation and Scattering of Waves", *IEEE Press*, New York, Second Edition.

Ivrissimtzis, L. P., and R. J. Marhefka [1992], "A uniform ray approximation of the scattering by polyhedral structures including higher terms," *IEEE Trans. on Antennas Propagat.*, AP-40, 1302-1312.

Keller, J. B. [1962], "Geometrical theory of diffraction", *J. Opt. Soc. Am.*, 52, 116-130.

Knott, E. F. and T. B. A. Senior [1973], "Equivalent currents for a ring discontinuity," *IEEE Trans. on Antennas and Propagat.*, AP-21, 693-695.

Kouyoumjian, R. G., and P. H. Pathak [1974], "A uniform geometrical theory of diffraction for an edge in a perfectly conducting surface," *Proc. IEEE*, 62, n.11, 1448-1461.

Kouyoumjian, R. G., and P. H. Pathak, [1986], "A uniform GTD approach to EM scattering and radiation," in Acoustic, Electromagnetic and Elastic Wave Scattering-High and Low Frequency Asymptotics, vol. II, *Varadan and Varadan, Eds. Amsterdam*, North Holland.

Lee, S. W. and G. A. Deschamps [1976], "A uniform asymptotic theory of EM diffraction by a curved wedge," *IEEE Trans. on Antennas and Propagat.*, AP-24, 25-34.

Maci, S., R. Tiberio, A. Toccafondi [1994], "Diffraction at a plane angular sector", *Jour. of Electromag. Wave and Appl.*, 8, 1247-1276.

Maci, S., R. Tiberio, A. Toccafondi [1996a], "Incremental diffraction coefficients for source and observation at finite distance from an edge" *IEEE Trans. on Antennas Propagat.*, AP-44.

Maci, S., P. Ya. Ufimtsev, R. Tiberio [1996b], "Equivalence between physical optics and aperture integration for open-ended waveguides" *IEEE AP-S/URSI, International Symposium*, Baltimore, Maryland. (Also submitted to IEEE Trans. on Antennas and Propagat.).

Michaeli, A., [1984], "Equivalent edge currents for arbitrary aspects of observation," *IEEE Trans. Antennas Propagat.*, AP-32, 252-258.

Michaeli, A., [1986], "Elimination of infinities in equivalent edge currents - I: Fringe current components," *IEEE Trans. Antennas Propagat.*, AP-34, 912-918.

Michaeli, A., [1995], "Incremental diffraction coefficients for the extended physical theory of diffraction," *IEEE Trans. Antennas Propagat.*, AP-43, 732-734.

Mitzner, K. M. [1974], "Incremental length diffraction coefficients," Technical Report No. AFAL-TR-73-296, Northrop Corporation, Aircraft Division.

Oodo, M. and M. Ando [1995], "Equivalence of physical optics and the aperture field integration method in the full pattern analysis. Effects of integration surface", *IEEE AP-S, International Symposium*, Newport Beach, California, 2, 889-892.

Pathak, P. H. [1992], "High-frequency techniques for antenna analysis," *Proc. IEEE*, 80, 44-65.

Ross, R. A., [1966] "Radar cross section of rectangular flat plates as a function of aspect angle," *IEEE Trans. on Antennas and Propagat.*, AP-14, 329-335.

Rubinowicz, A. [1965], "Darstellung der Sommerfeldeschen Beugunsgwelle in einer Gestalt, die Beitrage der einzelnen Elemente der beugende Kante zur gesamten Beugungswelle erkennen last". *Acta Phisica Polonica*, 28, fasc. 6 (12). 841-860.

Ryan, C. E., Jr. and L. Peters, Jr. [1969], "Evaluation of edge-diffracted fields including equivalent currents for the caustic regions," *IEEE Trans. Antennas Propagat.*, AP-17, 292-299. (see also correction, AP-18, p. 275, March 1970).

Sikta, F. A., W. Burnside, T. T. Chu, L. Peters Jr. [1983], "First-order equivalent current and corner diffraction scattering from flat plate structures," *IEEE Trans. Antennas Propagat.*, AP-31, 584-589.

Shore, R. A. and A. D. Yaghjian [1988], "Incremental diffraction coefficients for planar surfaces," *IEEE Trans. Antennas Propagat.*, AP-36, 55-70.

Tiberio, R. and S. Maci [1994], "An incremental theory of diffraction: scalar formulation", *IEEE Trans. Antennas Propagat.*, AP-42, 600,-612.

Tiberio, R., S. Maci, A. Toccafondi [1995], "An incremental theory of diffraction: electromagnetic formulation", *IEEE Trans. Antennas Propagat.*, AP-43, 87-96.

Ufimtsev, P. Y. [1971], "Method of edge waves in the physical theory of diffraction," translation prepared by the U.S. Air Force Foreign Technology Division, Wright-Patterson AFB, OH, released for public distribution on Sept. 7, 1971.

Ufimtsev, P. Ya. [1989] "Theory of acoustical edge waves," *J. Acoust. Soc. Am.*, 86, 463-474.

Ufimtsev, P. Ya. [1991], "Elementary edge waves and the physical theory of diffraction," *Electromagnetics*, 11, 125-159.

COMMUNICATIONS BY MEANS OF LOW EARTH ORBITING SATELLITES

Raymond L. Pickholtz

1 Introduction: Personal communication by satellite

Many organizations have proposed satellite-based personal communications providing global coverage for voice and data to hand-held subscriber units with a direct link to a satellite. Such systems will offer the ultimate promise of personal communications of allowing communications to the *person* at any time and in any place, literally, rather than to a physical location. The projected system capital costs are estimated variously as being between \$1 - 4 thousand million dollars. While satellite communications for trunked transoceanic telephone traffic has seen successful operation for more than a quarter of a century, the application of satellites to mobile communications applications and for direct broadcasting is a relatively recent development. To some extent, the change in focus was due as much to the "push" of competition from wideband digital, transoceanic optical fiber cable as to the "pull" of satellites being able to service a new market of mobile communications and, it is hoped, multimedia, nomadic personal communications, and computing. The unique opportunities that satellites have are for covering entire globe with minimal terrestrial infrastructure. Thus, initially, at least, satellites can be used for obtaining telephone and data communications in parts of the world where none exists at present.

Traditional communication satellites reside in a geosynchronous equatorial orbit (inclination 0°) at an altitude of 35,784 Km. While such systems offer significant advantages in terms of coverage per satellite and other benefits, this is not the typical approach being taken by current developers of mobile satellite communication systems in general, and for Personal Communications in particular. In the next section, we discuss, briefly, the issues that drive towards Low Earth Orbiting Satellites (LEOS).

2 LEOS, MEOS, and GEOS

During *WARC'92,* several frequency bands were established for mobile use on a worldwide basis. Since then, there have been numerous proposals for implementing such systems. There are proposals for using Geosynchronous Orbit Satellites (GEOS) at an altitude 35,784 Km, Medium Earth Orbit Satellites (MEOS) at 5,000-10,000 Km, and Low Earth Orbit Satellites (LEOS) at 150-1,500 Km. Of these, LEOS have attracted the most attention because of technical advantages and the novelty of having many satellites, handoffs, and a cellular-like configuration. The advantages include small propagation loss so that handsets could be used for direct communication from a mobile user, and small propagation delay (about 10 *ms* compared to 250 *ms* for GEOS) for better performance of voice, data, and other interactive services. In addition, LEOS do not suffer from consistent low elevation angles at high latitudes and the associated propagation anomalies that GEOS do. Disadvantages are that more satellites are required - offset by cheaper launch costs - and increased probability of shadowing and increased Doppler shifts.

LEOS orbits are placed beneath the lower of the two Van Allen radiation belts as shown in Figure 1 so as to minimize the radiation damage to electronic components that would result from a relatively unshielded, lightweight satellite. Extensive ionizing radiation severely reduces useful satellite life. On the other hand, such lower orbits experience slight, but greater atmospheric drag than higher orbits and, this too, reduces satellite life, thereby forcing smaller solar cell arrays and less primary power. LEOS satellites with significantly inclined or polar orbital constellations can easily cover higher latitudes and, to some extent therefore, even avoid blockage from tall structures.

There are additional major issues distinguishing LEOS and MEOS from GEOS. Some of the major issues are due to the fact that the former do not appear stationary to a ground user. Thus, the satellites in a LEO (MEO) orbit will pass overhead from horizon to horizon in a short time. This requires the implementation of a handoff mechanism and tracking. The additional design challenges include required worldwide coverage, maximizing system user capacity, allocating margins of fading, controlling interaction delay, call handoffs, spectrum sharing, maximizing handheld battery life, and restricting user health hazards (transmit power). The frequency bands to be used for LEOS is 1616.0-1626.5 MHz for the uplink and 2483.5-2600 MHz for the downlink if frequency division duplex (FDD) is used and just the former band if time division duplex (TDD) is used. Of five original proposals for "Big" LEOS, four have opted for CDMA in one form or another, and only one has proposed FDMA/TDMA. Because of this, and general interest in CDMA in LEOS, we have focused our effort in this paper on LEOS CDMA and, in particular, how it differs from terrestrial CDMA which has already demonstrated its virtues and has become one of several cellular standards in the United States and elsewhere.

As in cellular communications, maximizing area capacity (users/Km^2) requires a sufficient number of spot beams (cells) per satellite necessary and, to minimize interference, consideration must be given to frequency reuse patterns. The ephemeral visibility associated with low orbit means that, for LEOS and MEOS, continuous communication requires a "hand-off" procedure between spot beams and satellites, not unlike cellular terrestrial systems when the user moves from one base station coverage zone to another.

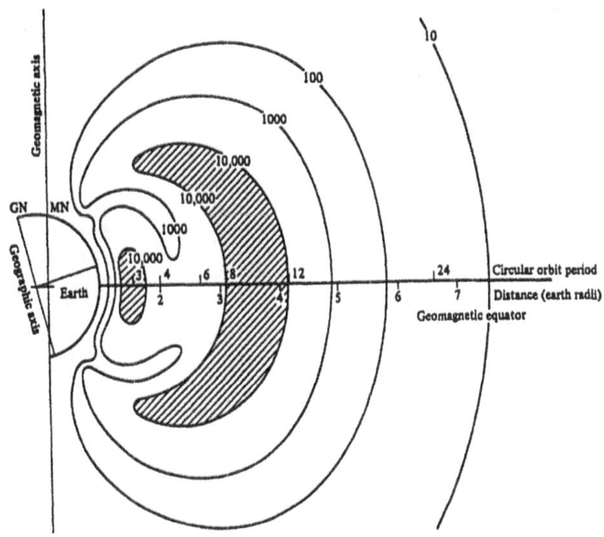

Figure 1: Relative flux levels of Van Allen radiation belts [Wu et al., 1994].

To understand the required number of satellites, N, for global coverage, we can consider multiple polar orbits of many satellite per orbital plane. In this case, the orbital planes of each satellite group within an orbit remains fixed (unlike inclined low orbits whose orbits drift due to the oblateness of the earth). We like to minimize $N = PS$, the total number of satellites to get full global coverage, where P is the number of polar planes and S is the number of satellites per plane. It is possible to solve this geometric optimization problem [*Werner et al.*, 1995] and

$$N \simeq \frac{4\sqrt{3}}{9}\left(\frac{\pi}{\gamma}\right)^2 \tag{1}$$

where γ is the earth central angle, i.e., the angle from the earth's center to the edge of a single satellite "footprint" circle on the earth. For example, the original proposed IRIDIUM constellation consisted of 77 satellites: $P = 7$, $S = 11$ yielding $\gamma = 18.44°$ and from this we can determine the minimum elevation angle on the ground. This established the satellite orbit altitude for a minimum elevation angle in the ground of 8.2°. Later, due to improvements in design, a slightly higher orbit was chosen so that $N = 66$ and $P = 6$, $S = 11$. For non-polar orbits, it may be possible to use a smaller number of satellites if certain regions of the earth are not covered as well.

LEOS have also been functionally subdivided as "Little LEOS" and "Big LEOS" with the Big LEOS offering wide band voice and data and the Little LEOS offering paging, short message services, utilities meter reading, inventory location, etc.

System (company)	No. satellites	Orbit inclination (altitude, Km)	Spot beams per satellite	Multiple access/duplexing processing	Launch date
IRIDIUM (Motorola et al.)	66	polar (780)	48	TDMA/TDD on-board demodulation, switching, cross links	1997
Globalstar (larel/Qualcomm)	48	52° (1,414)	16	CDMA/FDD "Bent pipe"	1998
Odyssey (TRW)	12	50° (10,345)	19	CDMA/FDD "Bent pipe"	1998
ICO (Immarsat)	12	45° (10,400)	?	TDMA/FDD "Bent pipe"	1999
LEONET (ESA)	12	54° (6390)	37	?	?

Table 1: Big MEOS/LEOS

The current status of the Big MEOS/LEOS for now (early 1996) is summarized in Table 1. Notice that only IRIDIUM has on-board processing with switching to other satellites via cross links. Also, only IRIDIUM uses TDD so that only one band (1616.0-1626.5) is used for both, up and down link communications.

Others, including Aries, Ellipso and AMSC have licenses deferred and may reenter. Some of these original systems also did not quite fit the MEO/LEO category described above.

There is an almost endless number of Little LEOS which will provide pager-sized two way short message personal communications services, direct satellite linkage to Laptop and Palmtop computers, remote telemetry, meter reading, and inventory tracking and location. These are all very low (10- 1000 bps) data rate services with time latency tolerances so that store and forward techniques can be employed. The frequency bands that have been assigned for this use are 399-401 MHz, 148-150 MHz, 137-138 MHz and a portion of the 2 GHz personal communications band. A representative set of Little LEOS is given in Table 2. The first launches are in the 1996-97 period.

System (company)	No. satellites	Inclination/altitude (Km)
Orbcomm (OCC)	36	70° & 45° /775
GemNet (CTA)	38	50° / 1,000
Starsys (SGPS)	24	53°/ 1,000
Faisat (FAI)	26	83° & 66° / 1,000
Leo One (LO USA)	48	50° /950

Table 2: Little LEOS

3 Design issues and architecture

The design of a complete system involves many choices and compromises. As we discussed above, the reasons for selecting LEOS for direct personal access to a satellite is largely dictated by low power requirement from hand-held subscriber units, and small propagation delays. However, in order to achieve maximum capacity-coverage, many satellites are deployed and, to maximize the spectral usage, multiple spot beams are formed by antenna arrays from each satellite. Therefore, the basic geometric unit on the ground is a spot (similar to that of a cell in terrestrial cellular system) with a diameter of a metropolitan trading area which moves with the satellite at 5,000 plus meters/sec depending on the orbit altitude. Therefore, even a high speed jet aircraft appears to be relatively stationary. The spots traverse the users rather than the users traversing the cells as they do in terrestrial cellular systems.

A major issue in multiuser communications design is the access method. There are many possibilities but matching commonly deployed commercial digital technology is a factor that mitigates in favor of just two choices at this time: TDMA and CDMA, where it is understood that each of these operates in a collection of possible subbands and thus have an FDMA component. For example, the IRIDIUM system FDM structure divides the available part of the 1616.0-1626.5 KHz band into 31.5 KHz subbands with guard bands sufficient to space the individual carriers 41.67 KHz apart. Within each 31.5 KHz subband, 50 Kbps QPSK is transmitted as 90 *ms* TDMA frames to form 4 uplink (UL) and 4 downlink (DL) channels by means of time slots as shown in Figure 2. Here the 90 *ms* TDMA frame is shown with the 4 duplex time-slots. There is a 0.1 *ms* guard time between ULl - UL2, UL3 - UL4, DL1 - DL2, and DL3 - DL4 otherwise interslot time guards are 0.28 *ms*.

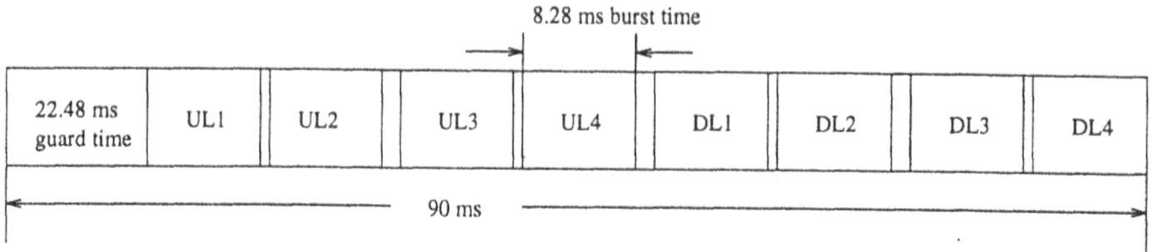

Figure 2: IRIDIUM TDMA frame structure.

In the CDMA systems, the basic structure is based on the terrestrial CDMA signal structure developed by Qualcomm and standardized by IS-95. It consists of 1.25 MHz FDM subbands and the duplexing is Frequency Division Duplexing (FDD) so that the 1.25 MHz subbands required per traffic channel are paired. The 1600 MHz band is used for the transmitting each 1.25 MHz uplink and the 2.4 MHz band is used for the corresponding 1.25 MHz in the downlink.

One of the major questions in the design is the number of spot beams per satellite and the so-called frequency reuse pattern. Similar to cellular systems, unless some means are taken to reduce the "spillover" of interference from adjacent spots (cells), the performance is unacceptable. As in cellular, when TDMA is used, this interference is controlled by not reusing the same frequency in adjacent spots. Typically, a 7 frequency reuse pattern is used whereby "central" spot is surrounded by 6 spots (approximated by hexagons) whose FDM frequencies are different for each spot in any cluster of seven spots as illustrated in Figure 3 by circles.

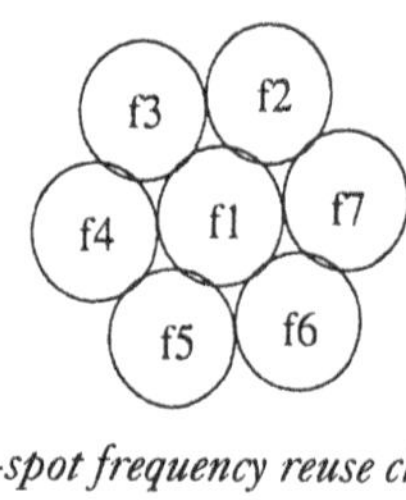

Figure 3: A 7-spot frequency reuse cluster in TDMA.

For CDMA, it is possible to have a frequency reuse pattern of 1. That is because of the spread spectrum processing gain attribute of CDMA. The processing gain mitigates multiple access interference (MAI) in CDMA by using a correlation receiver that discriminates between desired and undesired signals. Thus, the *same* set of frequencies may be reused in each (including any adjacent) spot. This factor (7 in this instance) shows up favorably to CDMA in computing the spectral efficiency (user channels/MHz). Offsetting, of course, is that each user signal spectrum must be spread to achieve the required processing gain. There are many other direct and some subtle issues that affect CDMA capacity in LEOS that we will address subsequently.

The architecture of a LEOS system is most strongly influenced by where intelligence for paging, call establishment, code/time-slot, and frequency assignment and hand-off signaling take place. Since CDMA processing is more intensive than TDMA, it makes sense to do most of that processing in a ground-based gateway and to keep the satellite very simple. Figure 4 illustrates the architecture of a "bent pipe" satellite network. Each satellite establishes a (moving) footprint which is in communication with a gateway shown here as the dish antennas. The individual ground users establish a traffic link to the satellite via a spot beam within the footprint (not shown). All such communications must go up (down) to (from) the gateway. For users in the terrestrial network or in another satellite footprint, terrestrial lines must be used to complete the circuit. The bent pipe approach is simple and inexpensive to build, but it depends on the terrestrial infrastructure for the networking, and many gateways are necessary for coverage. Most LEOS systems use this architecture.

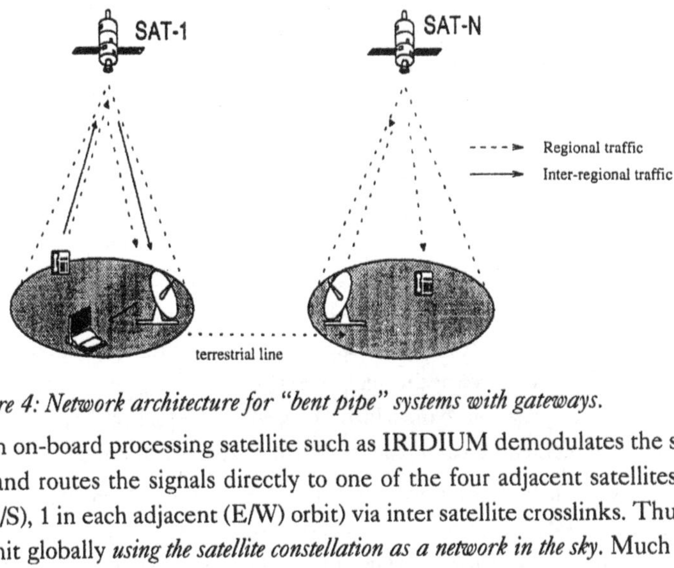

Figure 4: Network architecture for "bent pipe" systems with gateways.

By contrast, an on-board processing satellite such as IRIDIUM demodulates the signal, reads addresses, and routes the signals directly to one of the four adjacent satellites (2 in the same orbit (N/S), 1 in each adjacent (E/W) orbit) via inter satellite crosslinks. Thus, it is possible to transmit globally *using the satellite constellation as a network in the sky*. Much fewer gateways are needed. The gateways in both architectures provide for interconnection with the terrestrial networks such as the PSTN using switching systems and SS7 signaling. Because signals can traverse the major portion of a long distance connection via the satellite system in IRIDIUM, fewer long distance toll charges are incurred.

4 CDMA Capacity Issues- Contrast with Terrestrial

All cellular systems today achieve spectral efficiency measured in users/MHz/Km^2 by geographically selective reuse of subbands in the allocated spectrum. In both, conventional analog (AMPS) and digital (D-AMPS and GSM) systems, adjacent "cells" avoid using the same carrier frequency in adjacent cells as in Figure 3. For those cells sufflciently isolated electromagnetically by the attenuation of distances, terrain, buildings, and vegetation, the frequencies can be reused subject only to required mutual interference conditions (typically better than -18 dB). In terrestrial CDMA, it is usually possible to reuse all frequencies in each cell, including those that are adjacent. There are several reasons why this is possible in terrestrial CDMA. First, the nature of the spreading sequences are such that they exhibit low cross-correlation due to the spread ratio of $(1.25 \times 10^6)/(2 \times 4800) \simeq 130$ on the forward (gateway to user) link and (1.25 x 106)/(3 x 4800) $\simeq$ 87 on the reverse link. The voice codec rate assumed here is 4800 bps (this may be increased for better quality) and the factors of 2 and 3 in the denominators are the error correction code rates, respectively. This spreading ratio (processing gain) affords reduction in the mutual multiple access interference (MAI) and represents a rough measure of the maximum number of equal strength, mutually low cross-correlation users possible. Indeed, for the forward link, ignoring Doppler shift, the cross-correlation of signature (spreading) sequences are typically orthogonal so there is theoretically no interference from other users coming from the base station *within* the cell of the user.

Secondly, the signals from a transmitter at the surface of the earth attenuate close to an inverse *4th* power (rather than that of free space inverse *2nd* power) due to multipath and scattering because of terrain. This affords greater intercell isolation and reduced interference from signals at the same frequency. Thirdly, in many instances, discrete multipath spread is in the order of 5-10 μs (and up to 100 μs in mountainous areas). This effect can be exploited by the use of multiple, time spaced correlators (multiple-fingered RAKE) to achieve additional diversity gain. This latter effect requires that the reciprocal of the spreading bandwidth (1.25 MHz), equal to 0.8 μs is smaller than the delay spread, which is typically true in a terrestrial environment.

In general, fairly tight power control is required to maintain the full capacity promised by CDMA. This is necessary because CDMA is by nature a self-interference limited system. Thus, if for whatever reason (e.g., some users being closer than others to the reception/transmission source, unequal fading or shadowing, etc) the relative powers of the multiple user signals are unequal, they will use up more than their "quota" of interference and thereby reduce capacity. Indeed, accurate and tight power control is a major achievement in making CDMA for terrestrial applications work (typically achieving 1-2 dB power control error over a 90-100 dB dynamic range). In terrestrial systems, it is possible to use a closed-loop power control mechanism to achieve this because loop delays are relatively small (5-20 μs) so that the closed loop dynamics can track variations due to motion and fading. In satellites, the round-trip delay, even for LEOS, is about 10 *ms* or more and inhibits the use of tight closed loop power control. Therefore, the power control error is greater and this will manifest itself as a reduction in total capacity as we will show below. It might appear that open loop control can be used in the sense that the received signal strength is, by electromagnetic reciprocity, closely proportional to what needs to be transmitted and could be used to maintain equalized power levels. Unfortunately, even if this was true at a single frequency, CDMA LEOS systems have their up/downlink frequencies separated by 800 MHz. Only gross shadowing loss will be reciprocal; fading statistics and dynamics would be virtually independent in frequencies separated by 800 MHz. Furthermore, the 10 ms delay introduces another unrelated constraint and that is a restriction on the depth of interleaving for the error correcting codes. Deeper interleaving spreads the expected error burst so that channel noise and fading do not cause clusters of errors (or outright coding failure). Since the total allowable delay budget is limited for interactive services (like voice) and interleaving uses up this delay as does round-trip propagation, the interleaving depth must be restricted (usually to 20-40 *ms*). This problem is particularly acute in MEOS. If the interleaving depth does not cover typical fade durations, the decoding power is degraded and consequently results in a loss of capacity (since less MAI is tolerated).

In satellite systems, there is no *4th* power with distance from cell (spot) center. The interference from adjacent spots (and other satellites) on the ground is controlled entirely by the antenna pattern isolation. This also usually increases the MAI from adjacent spots compared with that of adjacent cells in terrestrial cellular systems.

Terrestrial CDMA and TDMA cellular use *sectorized* antennas whereby each cell site radiates three (or more) independent beams, 120° apart, and when increased traffic demands it, cell splitting is employed; that is, additional cell sites are deployed to create a finer cellular structure. These are particularly valuable techniques for CDMA since each sector and split cell (up to a limit) has *almost* the same capacity as the original cell without having to incur frequency reuse provided that the isolation of the sectors is sufficient. No additional rf carriers are needed to provide the additional capacity. Satellite systems do not enjoy this opportunity.

In LEOS/MEOS, there are higher Doppler and tracking requirements and there usually will be significant Doppler difference in satellite handoffs; thereby further complicating the reception problem. These handoffs are required in LEOS/MEOS even if the users are not moving. There are similar differences between terrestrial TDMA but their effects on capacity are not as pronounced.

For all LEOS/MEOS/GEOS systems, there are power spectral density limit set by *WARC'92*. For the uplink, it is -15 dBW/4 KHz and for the downlink there is a power flux density limit of -142 dBW/m^2/4 KHz as set in WARC'92 footnote 753F for high elevation angles in the 2483.5-2500 MHz band to protect terrestrial line-of-sight links. This is also covered in ITU regulation RR2566.

5 CDMA Capacity Degradations in LEOS

CDMA will experience performance degradations when operating in a LEOS compared to terrestrial due to the issues identified above. We reiterate here the major differences between the two environments so that we can examine how the loss of certain attributes in terrestrial CDMA that are not available to satellite CDMA degrades the latter's capacity.

Terrestrial cellular systems CDMA enjoys some considerable advantages.

- Capacity gain
  - Frequency reuse pattern = 1
  - Inverse 4*th* power adjacent cell interference
  - Voice activity gain
  - Sectorization gain
- Extended coverage and performance gain
  - Powerful coding gain is achievable with little penalty
  - Sectorization provides improved link budget from antenna gain
  - Soft handoff gain
- Multipath utilization
  - RAKE receiver (for multipath spread > 1 μs)
- Tight power control
  - Due to low round-trip delay Some of these advantages disappear in satellite operation and consequently there is a loss in capacity.

The usual approximate method of computing CDMA capacity within a spot starts with the equation [*Vojcic, Pickholtz,* and *Milstein,* 1994]

$$\mathrm{SNR} \simeq \frac{1}{\frac{No}{2E} + \frac{KI}{2L}\gamma} \tag{2}$$

where

SNR	=	the required signal to noise ratio,
E/N_0	=	energy per bit to white noise power spectral density density,
K	=	number of active users/spot,
I	=	ratio of the total multiple access interference (MAI) that a user experiences due to all spot beam to that experience by other users within the spot,

L = processing gain due to spread spectrum = W/R where W is the spreading bandwidth and R is the data rate,

γ = factor which takes into account voice activity (0.4-0.5) and the effect of shadowing. Strictly, this is a random variable which should be averaged.

The first term in the denominator is due to thermal noise and in a well designed system this should be small, leaving the SNR to be dominated by the second MAI term. Under these conditions (attempting to make K as large as possible), we get

$$K \simeq \frac{2L}{I\gamma} = \frac{2W}{I\gamma R} \frac{1}{(\mathrm{SNR})_{\mathrm{required}}} \tag{3}$$

as the number of users per spot.

In fact, (2) and (3) are fairly poor approximations when there is statistical shadowing and fading present. In satellite links where there is both, measurements [S] indicate that a good model is a mixed Rayleigh, Rician distribution with probability density function of the received amplitudes given by

$$f_R(R) = B\frac{R}{\sigma^2}\exp -\left(\frac{R}{2\sigma^2}\right) = (1-B)\frac{R}{\sigma^2}\exp -\left(\frac{R^2+A_s^2}{2\sigma^2}\right).I_0\frac{RA_s}{\sigma^2} \tag{4}$$

where A_s is the amplitude of the specular component of the Rician part (second term), $2\sigma^2$ is the average power in the scatter component, $I_0(\cdot)$ is the modified Bessel function, and B is the probability of showing (fraction of shadowed user). After we normalize the parameters $A_s = 1$, $c \triangleq 1/2\sigma^2$, we can compute the *average* SNR which includes not only the effects of shadowing and fading, but an effect similar to fading but independent of it; i.e., the effect of imperfect power control is equivalent to the received signal amplitude being a random variable. Furthermore, we can incorporate the gains of error correcting coding and the loss in some of that gain due to imperfect interleaving, and finally identify what gains dual satellite diversity achieve for comparison. An example of a set of such calculations [*Vojcic, et al.*, 1994-1; *Vojcic, et al.*, 1994-2] is shown in Figure 5.

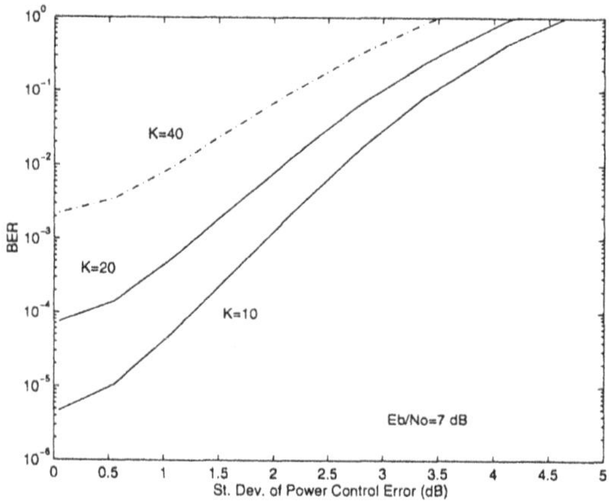

Figure 5: Effects of imperfect power~control.

This plot shows the rapid degradation of performance with increased power control error standard deviation. For example, if one requires a bit error rate of 10^{-3} (voice codec), then for the parameters of this plot, a 2 dB power control error (typical for terrestrial closed loop system) will allow only 10 users per spot (K). Satellite system power control error is expected to be larger than this for the reasons discussed above. In this plot, we are examining the shadowed user but average the MAI over shadowed and non-shadowed users. Rate 1/3 convolutional coding is included. *Lutz et al.*, [1991], show that for the same parameters, dual satellite diversity allows quadrupling of the number of users per spot, but that requires that two satellites always be in the user's view. These considerations, especially when there is additional loss from imperfect interleaving, leads to the conclusion that satellite diversity is necessary to maintain capacity. Of course, with CDMA this diversity is easily achieved with a RAKE receiver provided that both satellites are present, transponding (bent pipe) the same signal delayed by at least $1\mu s$.

6 Bandsharing in CDMA

A major issue that has emerged in LEOS system is that there are many players and a limited amount of assigned bandwidth currently. *WARC'92* assigned 1610-1626.5 MHz (L band) and 2483.5-2500 MHz (S band) to LEOS services on a *worldwide, primary basis*. While there are plans to release additional spectrum in the 1980-2010 MHz and 2170-2200 MHz range, all of this would still be insufficient to satisfy the many operators that have clearly gone a long way towards development and others that are planning to do so. In the United States, the *FCC* sought to mediate prospective spectral disputes by setting up a negotiated rule making (NRM) committee in 1993 whereby the parties, themselves, would resolve the question. There were basically two points of view. One, held by Motorola, the only TDMA system proposer (IRIDIUM) suggested that since S-band was filled with interference from industrial, scientific and medical (ISM) equipment that it would use only the L-band for both up and downlink using TDD. They then argued that this band be "segmented" into two equal 8.25 MHz bands, one for TDMA (IRIDIUM) and the other for all the other (at that time, 5) prospective CDMA service providers. The CDMA proponents argued that CDMA, being an "interference sharing" system (as described in equation (2)) would be able to *simultaneously* share the *entire* band. Indeed, the CDMA proponents argued that not only can they share the full band, but that, because of ITU RR2566 which specifies a maximum downlink power flux density (PFD) *per satellite* of -142 dBW/m^2/4 KHz that total band sharing actually results in a *greater* total capacity in CDMA than individual operators using a piece of segmented spectrum!

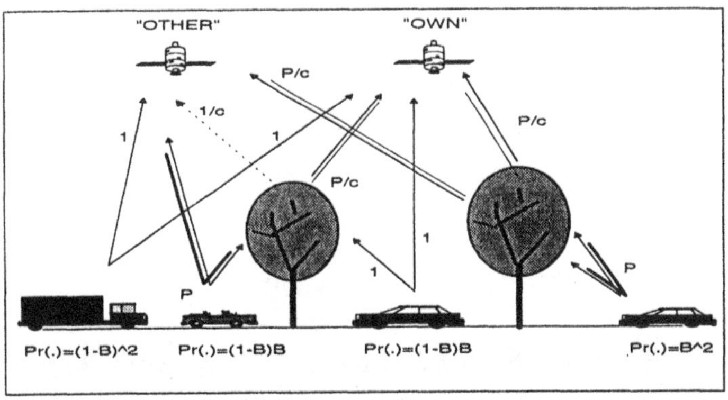

Figure 6: Multiple systems uplink sharing.

The reason for this apparent anomaly is not difficult to understand if one looks at the simple, approximate equation (2). If, because of PFD limits, the second term is upper bounded, i.e., $KIP < P_{max}$, where P is the power of a single user and the left part of inequality is proportional to the power flux density laid down by a *single* satellite. Then, the condition for achieving the ideal (for CDMA) of being interference (*2nd* term) limited is not realized for a single satellite and other operators can "lay over" the some band and use up the available interference budget! This situation, although plausible on paper with the simplified model of only interference and additive white Gaussian noise as implied by (2) is however, not that simple. In fact, if we add fading and shadowing using the channel model described in equation (4) the scenario that must be analyzed is shown in Figure 6 for the uplink. A similar, but not identical scenario appears in the downlink.

In this diagram we have two operators' satellite, "OWN" and "OTHER". A user on the ground who experiences shadowing (with probability, *B*) has a shadowing loss *l/c*, and its signal is largely Rayleigh faded and must be compensated for by additional power (exercised by power control) $P/c > 1$. However this additional power is simultaneously experienced by both, the "OWN" operator's non-shadowed users *and* by the "OTHER" operator's nonshadowed users as additional interference. While "OWN" might consider this additional interference as "sharing", the "OTHER" sees this as power "robing". The detailed analysis of this and other scenarios, including the downlink are given in [*Vojcic, et al.*, 1995]. An example of the results of the analysis is shown in Figure 7.

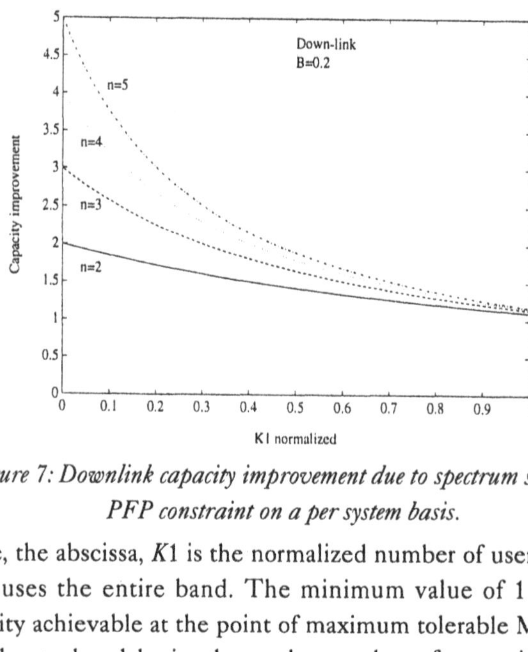

Figure 7: Downlink capacity improvement due to spectrum sharing-PFP constraint on a per system basis.

In this figure, the abscissa, *K*1 is the normalized number of users per spot beam if a single operator uses the entire band. The minimum value of 1 corresponds to the maximum capacity achievable at the point of maximum tolerable MAI. The ordinate is the multiplier due to bandsharing by *n*, the number of operations (used here as a parameter). The shadowing probability in this figure is taken at 0.2 and we can see that *in this case* where the PFD limit is on *per system* (satellite) basis. Then, indeed, we have greater total capacity when *n*, the number of operations is larger than 1. However the greater gain occurs when *K*1 is very small. That is, when each system has very little capacity to begin with! Furthermore, since *K*1 is monotonic with the power flux density limit, for the parameters in the figure the RR2566 PFD limit corresponds to a *K*1 of above 0.8. At that point there is very little gain in multiple operator band sharing.

Indeed, if one is realistic about the purpose of the PFD limit, one should impose a PFD on the *aggregate* of *all* satellites since the objective is to limit total interference caused by satellites to other services on the ground. When this is examined, the gain due to power sharing virtually disappears even at low PFD as shown in Figure 8.

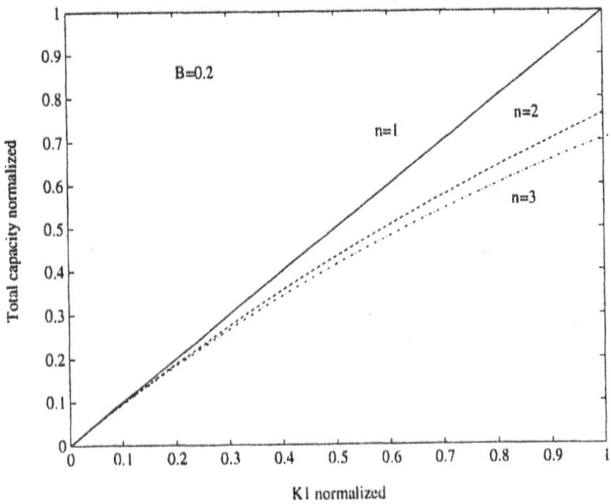

Figure 8: Downlink capacity improvement due to spectrum sharing- aggregate PFP constraint.

Indeed, here we see a *reduction* in total capacity due to sharing. The results carry over qualitatively even with dual satellite diversity. These results are based on the most optimistic assumptions which includes perfect power control *across all operators*, all satellites operating at approximately the same altitude and power level, etc.

At this point in time, the issue of CDMA sharing is somewhat moot since the NRM resulted in a ruling to segment the band and segregate the TDMA operators from the CDMA operators (now only two). Allocating the top 5.15 MHz of the 1610-1626.5 MHz band to TDMA and the bottom 11.35 MHz to be shared by *all* CDMA operators providing there are at least two. If there remains only one CDMA operator then CDMA and TDMA would each receive one half the band, a 8.25 MHz. There are, of course other requirements for the operators including continuous voice coverage of the globe for 75% of the time and 100% over the CONUS, assurances of lack of interference with radio astronomy and the need for K_a -band feeder link spectrum for the communications to gateways.

7 Summary and Comments

The application of LEOS to global personal communications is still in an embryonic stage. It is clear that there are many benefits to be derived by translating technology from terrestrial cellular systems to LEOS. However, we must be careful to note that there are substantial differences which, if not properly understood, will lead to disappointing performance and capacity. That said, there are many untapped techniques that are, with good justification, speculated to improve both performance and capacity dramatically. Among these techniques are adaptive array processing, multiuser detection methods, improved voice and video compression, more powerful error coding, and software radios.

Ultimately, however, the global nature of LEOS and its enormous coverage will require additional radio spectrum for traffic channels as well as for feeder links and cross-links. This pressure for more bandwidth will intensify as prices drop and the subscriber growth curve accelerates as it has done over the past few years for cellular.

Much more experimental propagation data is needed in the currently assigned and future LEOS bands if we are take maximum advantage of the signal processing techniques. This data needs to include not only signal strength, fading, shadowing and scattering statistics, but also time, frequency and space correlation characteristics. This would allow the formulation of channel models for the analytical and simulation evaluation of the techniques mentioned above.

Finally, some way needs to be found to allow currently incompatible standards to interoperate. Software radios may be able to make for some degree of operability of the air interface. Network interoperability may also be achievable using gateways with protocol conversion not unlike what terrestrial data network as the Internet has done. There is much to be done by the Radio Science community and there are few more exciting incentives to make an impact than improving LEOS systems for global communications.

References

FCC [1993], "FCC Document on NRM for MSS above 1 GHz", Washington, DC.

Lutz E. et al. [1991], "The Land Mobile Satellite Communications Channel-Recording, Statistics, and Channel Model", *IEEE Trans. on Veh. Technol.*, Vol. 40, No. 2, pp. 375-386.

Vojcic B. R., Pickholtz R. L., and Milstein L. B. [1994], "Performance of DS-CDMA with Imperfect Power Control Operating over LEOS", *IEEE J. SAC.*, Vol. 12, pp. 580-587.

Vojcic B. R., Pickholtz R. L., and Milstein L. B. [1994], "Downlink Performance of DS CDMA Performance over a Mobile LEOS Channel", *IEEE Globecom*, pp. 330-337.

Vojcic B. R., Milstein L. B. and Pickholtz R. L., [1995], "Total Capacity in a Shared LEOS Environment", *IEEE J. SAC.*, Vol. 13, No. 2, pp. 232-245.

WARC'92 [1992], *World Administrative Radio Conference*, Torremolinos, Spain.

Werner M., Jahn A., Lutz E., and Botcher A. [1995], "Analysis of System Parameters for LEO/ICO - Satellite Communication Networks", *IEEE J. SAC.*, Vol. 13, No. 2, pp. 371-381.

Wu W. W., Miller E. F., Pritchard W. L., and Pickholtz R. L. [1994], "Mobile Satellite Communications", *Proc. IEEE*, Vol. 82, No. 9, pp. 1431-1448.

TOPOLOGY BASED MODELLING OF VERY LARGE SYSTEMS

Jean Philippe Parmantier and Pierre Degauque

1 Introduction: Electromagnetic Topology, an appropriate conception method for modern EMC

Nowadays, it is obvious that Electromagnetic Compatibility (EMC) problems have to be taken into account at the conception phase of large systems. This is the best way to optimise the allocation of protection on the whole system. Nevertheless the solution dealing with shielding all the walls and systematically applying protection devices at the input points of equipment leads to a harmful increase of the weight of the systems. Besides, this solution makes the systems more complex to maintain in their operational life.

In the beginning, Electromagnetic Topology (EMT) could not be considered as a method to predict Electromagnetic (EM) interference. It arose in the 1980s as a support for the design of hardened systems against EM Pulse (EMP) [*Lee*, 1980]. The theory relies on decomposition of the geometry of the problem in different volumes included within each other (see Fig. 1). This breakdown is made possible thanks to the "Good Shielding Approximation" principle, which supposes that inner volumes cannot interact with external volumes [*Baum*, 1989]. In practise, such a breakdown is not realistic. Indeed, wires are inevitably present in systems, and make the volumes "electromagnetically" dependent. This is the main reason why, until 1990, the method remained at a conceptual stage and did not benefit from notable applications. Consequently, the objective of this article is to show that actual progress has been made on this topic in the last few years.

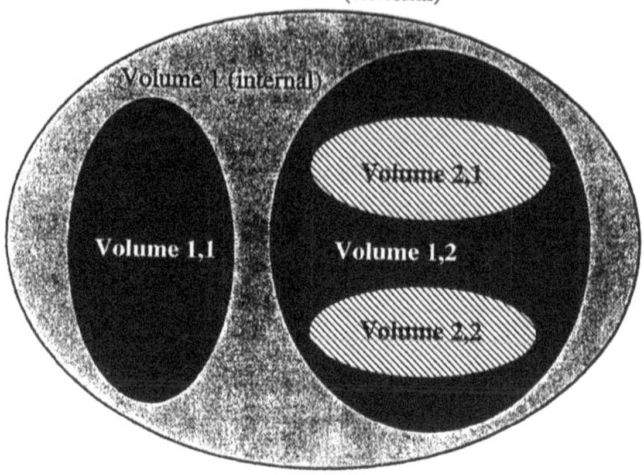

Figure 1: Example of a topological decomposition (topological diagram).

In 1988, the method was taken up again at ONERA, Dassault Aviation and Lille University [*Parmantier et al.*, 1989], with the objective being to adapt the method to obtain quantitative predictive results on practical EMC problems on large systems such as aircraft, satellites, and rockets. The spirit has been kept in mind because the geometric breakdown remains key to the method. From 1991, ONERA was in measure to demonstrate the interest of this method with some validations of the first version of its CRIPTE numerical code on scale models [*Parmantier*, 1991].

2 EMT versus classical EMC methods

EMT presents itself as an original method for efficiently integrating and controlling EMC constraints on a large system [*Baum*, 1982]. For this, it reveals itself as a quantitative tool to help in the prediction of internal interference. The need for EMT lies on the statement that a unique method (numerical or experimental) is not able to give significant results on complex systems. It is the reason why it can be said that EMT, because of the organisation of problems it offers, is an alternative to the two technical methods which are often proposed to deal with EMC problems: the large scale model experiments and the three dimensional numerical codes (3D codes).

Some years ago, experiments on scale one models were very frequent in helping the EM design of systems. When the exciting source could be synthesized correctly, these models gave very helpful results for EMC problems. Nevertheless, these problems could not be representative of other problems which would be encountered on future serial objects. Anyway, with the cost of programs becoming more and more expensive, most industrial companies have now to reduce the EM experimental part in programs and, generally, EMC tests are directly performed on prototypes, far after the conception phase.

In parallel to experiments, significant progress in computers (increase of memory storage capabilities, decrease of computation time), has opened new perspectives to 3D codes in EMC. The principle of 3D codes is to apply Maxwell's equation by meshing the entire geometry of the problem. In low frequency range (up to 500 MHz), these codes have proven to be perfectly adapted for the treatment of what is called the "external" problem. This means that, supposing the source is outside the system, surface current distribution and outer scattered field can be computed with great precision. But, the treatment of the "internal" problem requires a very precise description of the geometry. Particularly, despite the fact that most 3D computational technical methods are able to take into account single wires, they are not able to consider multiconductor type cables running inside the structure. To solve both external and internal problems with the same 3D code, one would have to mesh the geometry at the scale of the wires !

3 Specificity of electromagnetic topology

One of the limitations of both experimental and 3D numerical approaches deals in the fact that they try to solve the entire problem with a single method. EMT, however, does its best to decompose the entire problem in elementary problems, limited to sub-volumes having a lower size than the global problem for which a more appropriate method can be applied.

Notwithstanding the final objective, which is to treat the entire problem, EMT's objective is to make the treatment the most modular possible. A notable consequence is that all the work performed during the last five years in EMT showed that it was possible to introduce all the knowledge acquired in the past for the treatment of canonical problems [*Vance*, 1978; *Degauque and Hamelin*, 1990] to take into account EM coupling on complex systems. Besides, the modularity offers interests other than the capability to treat large scale problems. The consequences of modularity are numerous and can be summarised in three topics:

a- the hybridization of methods: thanks to the decomposition of the entire problem in elementary sub-problems, each one can be studied with a specific method. For example, one could analyse a particular sub-volume with a 3D code. Because the size of the geometrical problem has become smaller than the entire problem, one could benefit by all the power of its computer (either by reducing the computation time, or by having a more precise description of the internal geometry). With the same objective, one could also model a volume experimentally (for example, this is the case of electrical circuits which can be entirely characterized with measurements).

b- the parametrical approach: each volume can be studied separately of its neighbours, so it becomes easy to modify the characteristics in one volume, without calculating the characteristics of the other volumes (both the geometry or the electrical characteristics of this volume may be modified). In the same way, sources, can be one after the other switched in each volume: this is one efficient way to detect which sources are relevant for the system's immunity. On the contrary, with the global problem, the parametric approach is more difficult in 3D code calculations or in experimental approaches. It is the reason why both technical methods are generally applied on the more representative coupling configuration. Nevertheless, the main problem remains to detect what source is relevant in the EM interference for the correct behaviour of the system.

c- the use of "topological" data-banks: as often as possible, the topological method tries hard to characterize the constitutive elements of the problem under study independently of the structure. This is the case for elements such as cables, electrical circuits, current injectors, which can be characterized outside the structure, in a laboratory, and whose characteristics can be used in different problems.

To summarize, the topological approach on large scale structures is more a way of thinking whose objectives are three:

a- the organisation of EM coupling problems,

b- the research of the information in its simplest and most accessible form,

c- the will of using results obtained in other configurations.

4 Description of a topological approach

4.1 Decomposition in sub-problems

This operation is generally intuitive. The frontiers of sub volumes are very often constituted by physical surfaces (walls, boxes, cables). Nevertheless, two kinds of surfaces can be distinguished:

a - surfaces which are actual shields in the sense that they attenuate the signal from one side to the other side [*Clark and Baum*, 1988]: this is the case of the fuselage of an aircraft or a cables shield. They are called "proper" surfaces. These surfaces contain one volume called a "proper" volume. The main interest of detecting or building such surfaces is that they allow the external coupling and the interference induced inside to be treated separately. This comes from the application of the "Good Shielding Approximation", one of the basis of the theory [*Lee*, 1980],

b- surfaces which do not provide any kind of shielding effect, but which separate two physical volumes. This is the case of walls with apertures or cables crossing them [*Karlson*, 1978]. Consequently, the signal inside the separate volumes are dependent on each other. Such surfaces are called "elementary" surfaces and, in the same way, the included volumes are called "elementary" volumes. The interest is that these surfaces constitute a natural frontier for the decomposition and a place where observable data are easy to access. On a topological diagram, Fig. 2 depicts an illustration of both types of surfaces and volumes. This example, with a low number of proper surfaces and many elementary surfaces is more representative of an actual system topology than the one depicted on Fig. 1, [*Noss*, 1983; *Vétri and Costache*, 1991].

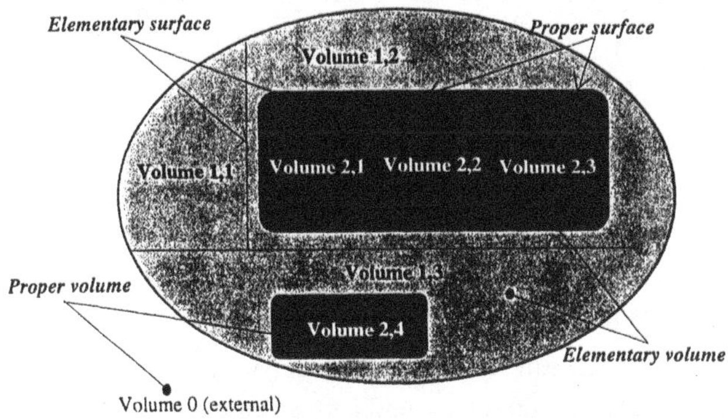

Figure 2: Terminology of surfaces and volumes in EMT.

4.2 Description of the global problem with the topological network

Then, following the volume decomposition, EMT proposes a network method to help to establish the solution for the global problem, thanks to the topological characterization of the scattering in each elementary problem.

The network can be seen as a formulation of the coupling existing between the different volumes which compose the entire volume under study. In the network, each volume (proper or elementary) is described in terms of a **junction**. Each junction is related to another junction thanks to a **tube**. The coupling of sources on the network is expressed in terms of equivalent generators applied on the tubes. Fig. 3 gives an example of a topological network and, especially, of the terminology used in CRIPTE (numerical code developed at ONERA to help on EMT treatments). Junctions may represent elementary elements such as circuits, but, more generally, junctions may represent a whole volume. This volume can be itself described by a network. Such a junction is called an "**equivalent junction**" [*Parmantier et al.*, 1992]. From their sides, tubes support the equivalent sources applied on the network. Particularly, they provide the "**inter-connections**" between "local networks" (see Fig. 3). A very important aspect is also that tubes support the possible propagation of signals between two junctions: in this way the tube is perfectly adapted to describe the particularly important problem of EM coupling on cables.

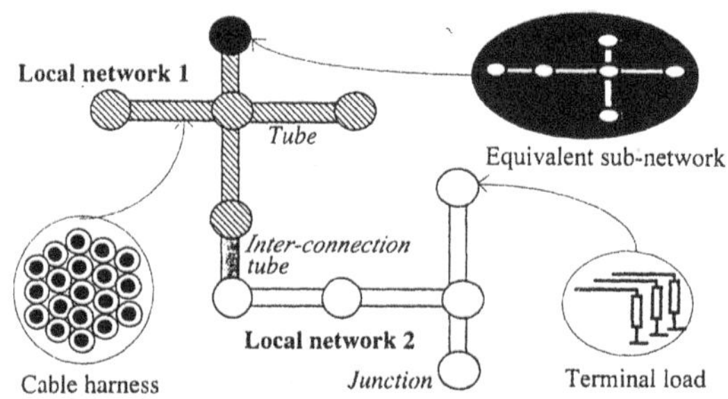

Figure 3: Topographical network description (CRIPTE code formalism).

4.3 Equations of the network

4.3.1 Scattering equation on junctions

At each junction (which can be seen as the separation frontier between several volumes), observable data are defined on ports. One can distinguish two kinds of ports:

a- ports which allow boundary conditions, compatible with the neighbourhood junctions. These observable data and ports depend on the type of sub-volume associated to the junction. If the junction represents a circuit, these ports are the external pins of the circuit and the observable data are voltage and current on them. If the junction is an actual volume, the ports may be associated to the fictitious cells used by a 3D code. In this case, the observable data can be the field distribution on an aperture, the current and voltage on wires.

b- ports on which one wants to compute a relevant signal. Generally these ports are made of wire extremities or pins on circuits, at the entry points of equipment.

All kinds of observable data can be expressed in terms of input waves W+ and output waves W-, combining two observable data on the ports. The wave definition depends on the type of observable data on the ports (for example, electric field and magnetic field, or voltage and current). The principle of characterization of the junction deals with expressing the coupling between W- and W+ on the different ports of the junction in terms of a **scattering matrix S.** The terms contained in the matrix are called "Scattering parameters" **(S-parameters)**. This scattering matrix summarizes all the scattering information of the volume at the level of the ports [*Parmantier et al.*, 1992]. The scattering equation at each junction can then be written:

$$W- = S\, W+ \qquad (1)$$

If the equivalent volume associated to the junction contains sources, these internal sources give rise to equivalent sources (also called **"Thevenin" equivalent generators)** at the level of each port of the equivalent junction.

4.3.2 Propagation equation on tubes

The tube, providing the connection of two junctions, can also be the support of signal propagation between two junctions and, consequently, the support of equivalent sources. In the general case, the source terms come from the junctions, and they can also be expressed in terms of source waves, Ws. Assuming a direction of propagation on the tube, the propagation equation is defined thanks to a **propagation matrix** Γ which relates waves W(0) and W(L) at each extremity of the tube. The definition is the following:

$$W(L) = \Gamma.W(0) + Ws \qquad (2)$$

If there is no effective propagation on the tube, Γ is equal to the identity matrix and the tube simply provides the connection of input and output waves of two junctions, thanks to a source wave excitation.

4.3.3 The BLT equation

The equation of the coupling on the network comes from the generalization of scattering and propagation equations on the whole network on which supermatrices (matrices made of matrices) and supervectors (vectors made of vectors) are defined. With this generalization, [W(0)] represents the output wave supervector and [W(L)], the input waves supervector at all the junctions. The BLT equation (from Baum, Liu and Tesche [*Lee*, 1980; *Baum*, 1989] allows to determine [W(0)] and takes into account the coupling on the whole network. The BLT equation is given by:

$$([I] - [S][\Gamma]).[W(0)]=[S].[Ws] \qquad (3)$$

where [I], [S] and [Γ] are, respectively, the identity, the network scattering supermatrix and the network propagation supermatrix and [Ws] the network source supervector.

4.3.4 Particular case of cable networks

In the frequency range from DC to 500 MHz, wiring plays a particularly important role in EMC problems because it drives the interference up to the entry of equipment. Consequently, for many years, a great amount of fundamental works has been concentrated in this frequency range which includes relevant threats such as lightening and EMP. From the EMT point of view, wiring is also a domain for which the network formalism seems very well appropriate [*Baum et al.*, 1988]. Indeed, to treat coupling on cables, the topological network formalism can directly be inspired from the multiconductor transmission line formalism. This formalism allows to model complex, shielded or unshielded, wiring such as cable harnesses, as they can be encountered in actual systems. Especially, the mutual coupling between wires and the different propagation velocities can be taken into account. The main specificity of multiconductor line network formalism lies in the notion of tube. The tube is physically associated to a piece of wiring for which the propagation characteristics remain identical on all its length (characteristic impedance Zc and propagation coefficient matrices γ). These characteristics also impose the definition of waves defined as a function of distributed current I(z) and voltage V(z) along the line:

$$W(z) = V(z) + Zc\, I(z) \qquad (4)$$

The other specificity deals in source waves which can be expressed as a function of current generators Is(z) and voltage generators Vs(z) distributed all along the tube:

$$Ws = \int_0^L \left(Vs(z) + ZcI(z)\right)e^{-\gamma(L-z)}dz \qquad (5)$$

In the multiconductor line network formalism, junctions generally represent the terminal equipment or the connection between tubes which have different characteristics or more generally, a junction equivalent to another network. The consequence on the scattering matrices is that they become referenced to the characteristics impedance of the tubes connected to it. The scattering parameters are then called "**topological**" scattering parameters. From them, the input impedance or admittance matrices of the junction can be easily computed [*Lee,* 1980; *Parmantier et al.*, 1992; *Besnier and Degauque*, 1995].

The tube characteristic matrices can be derived from the knowledge of the distributed impedance Z and admittance matrices Y. These matrices can also be expressed in terms of **resistance**, **inductance**, **conductance** and **capacitance** (R, L, G, C) distributed matrices. The important aspect is that these matrices can be determined experimentally or numerically, on canonical geometry (such as over a metallic ground plane), outside the structure under study. In this way, one can take advantage of keeping these parameters in a wiring data bank, and use them in other problems.

Since 1990, most of works performed in EMT dealt with coupling on wiring systems and a great experience has been acquired for the modelling of complex wire harnesses. The cable models can be derived from measurements [*Agrawal et al.*, 1979] or from two dimensional code computations (calculating Laplace's equation), but in both cases, a statistical treatment has to be applied to take into account the uncertain running of wires inside the harness [*Parmantier*, 1993]. The multiconductor line formalism has also proved its efficiency to model shielded cables, and especially to model shield connection imperfections [*Parmantier et al.*, 1993].

5 Applying the decomposition on topological networks

In fact, all EMC problems on systems can be described with a network. But the size of the entire network may still remain very large for computers. How is it possible to reduce the size of the network? The answer deals with using the properties of proper and elementary surfaces.

5.1 Applying breaking down on proper volumes

We saw previously that the interest of proper volumes was to allow the application of Good-Shielding Approximation. In fact, this concept is equivalent to say that the external volume (that is to say the volume supporting the source) generates equivalent generator distributions on the proper surface. These generators will then be applied to the internal volume. Each problem can then be treated by designing an associated topological network for which junctions represent the elementary volumes.

The following example shows how to use the topological breakdown of a problem to help to the design of the topological networks: it deals with an external EM coupling on an aeroplane. For the purpose of the demonstration, the topology of the aeroplane has been simplified (see Fig. 4). This example was the subject of several experimental works performed, on a scale model, at ONERA and Lille University, to validate the topological concepts [*Parmantier*, 1991; *Parmantier and Aparicio*, 1991; *Besnier*, 1993; *Besnier and Degauque*, 1995]).

The aircraft is made of four physical volumes:

- the right wing volume,

- the left wing volume,
- the fuselage volume,
- the medium volume, which connects the three previous volumes.

We have chosen to study the EM coupling on two canonical wiring paths:

- one connecting both extremities of the wings
- one connecting the extremity of the left wing to the nose of the fuselage.

Fig. 4 also presents three types of representative penetration paths of energy inside a structure:

- the apertures (apertures 1 and 2)
- the antennas (antenna HF and RADAR)
- the diffusion through walls (in the medium volume supposed bad conductor).

The first step of a topological treatment deals with the transcription of the geometrical analysis of the system in a topological diagram (see Fig. 5). This diagram shows two proper surfaces including two proper volumes:

- the external surface of the aeroplane (first shielding level)
- the surface of cables (second shielding level).

Each proper volume is composed of elementary volumes limited by the four physical volumes. One can distinguish:

- elementary volumes dealing with the internal space included in the first shielding level (V1,1 to V1,6); V1,5 and V1,6 represent in Fig. 5 the antenna feeders,
- elementary volumes dealing with the wiring in the four physical volumes, included in the second shielding level (V2,1 to V2,10).

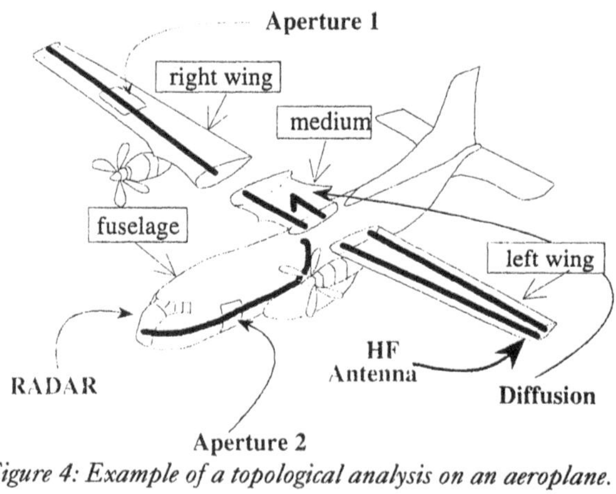

Figure 4: Example of a topological analysis on an aeroplane.

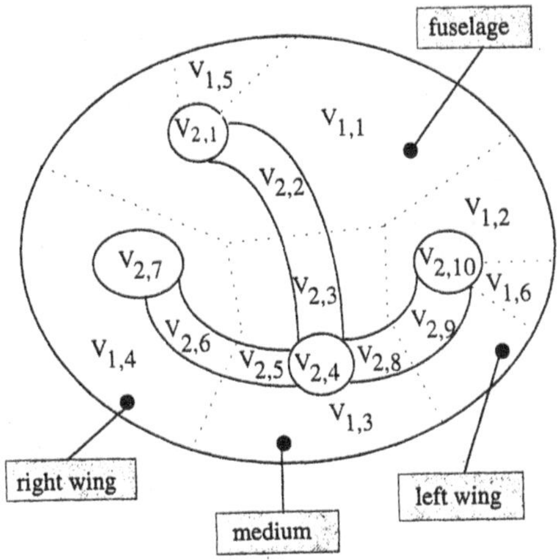

Figure 5: Decomposition of the aeroplane in a topological diagram.

From a quantitative point of view, a global decomposition of the problem into three steps, each one associated to the resolution of a network describing a proper volume, can be proposed:

Step 1: equivalent sources are calculated on the first shielding level, where energy penetrates inside the structure (apertures, running of cables from the outside to the inside). So, the topological network deals with a single junction whose ports represent the penetration paths.

Step 2: at low frequencies (typically under 500 MHz), one can take advantage of treating the wiring network as a third shielding level, on which Good Shielding Approximation can be applied. Equivalent sources calculated at the previous steps are used to derive the incident fields in the internal volumes, when wires are removed. Let us insist on the fact that these fields are "incident" fields, that means that, in the vicinity of cables, they can be totally different of fields in presence of the wiring. In the case of the aeroplane, the associated topological network is depicted in Fig. 6. Tubes are shrunk to zero length because junctions represent physical volumes.

Step 3: the response of the wiring is computed by injecting the incident fields as equivalent generators. For the case of the aeroplane, the network is depicted in Fig. 7. It is directly derived from the topology of the internal wiring. For this shielding level, tubes are associated to physical sections of wiring.

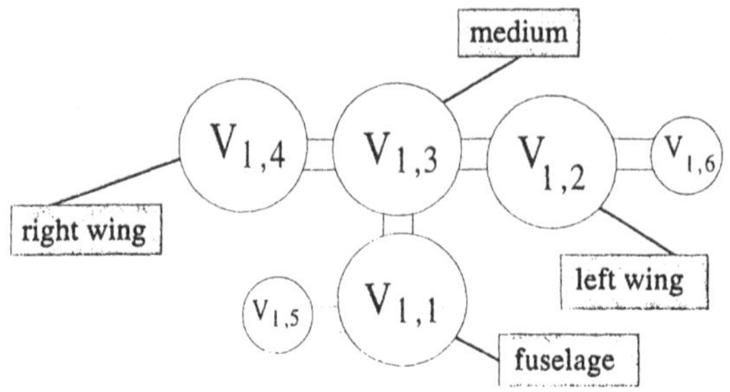

Figure 6: Topological network associated to the first shielding level.

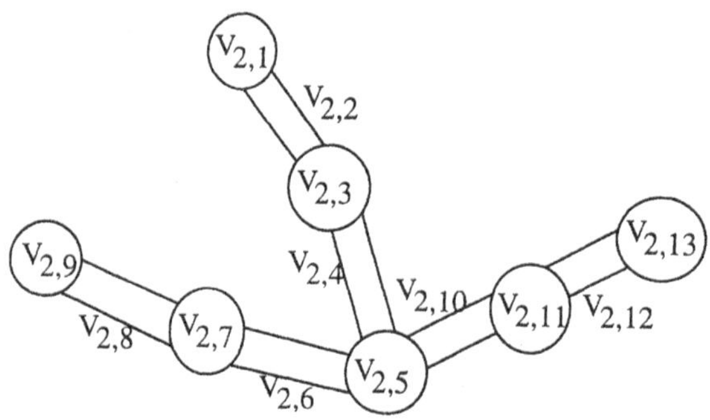

Figure 7: Topographical network associated to the second shielding level.

5.2 Practical breaking down on proper volumes.

In practice, up to 500 MHz, the three previous steps may be combined in different ways:

a- if a 3D code is used, step 1 and step 2 may be computed simultaneously by meshing the structure without the cables and deriving incident fields outside and inside the structure,

b- in other cases, only steps 1 and 3 may be computed if one finds how to express directly the coupling of the external equivalent sources on internal wires with transfer functions.

The following example gives an example of a case where step 2 is not necessary to perform. This example also illustrates the capability that EMT offers for coupling 3D codes with cable network codes. It comes from series of experimental campaigns on a modular caisson performed at Centre d'Etudes de Gramat (CEG) as a collaboration between ONERA and CEG from 1991 to 1992 [*Gobin et al.*, 1992]. The caisson under study was made of five elementary volumes, linked to each other by wires, whose running was complex enough for the problem not to be able to solve directly with a 3D code. The results presented here deal with the prediction of internal interference when the caisson is excited by an EMP shape wave delivered under CEG's SSR. (Simulator Semi Rhombic) and when the main penetration path is provided by an aperture in one of the five volumes. The coupling on wires running under a small aperture can be modelled by a voltage generator Vs and a current generator Is, given by the following expressions [*Lee*, 1980; *Yang and Baum*, 1986]:

$$Vs = j\,\omega\,\alpha\,Ecc \tag{6}$$

$$Is = j\,\omega\,\beta\,Hcc \tag{7}$$

The α and β transfer functions are related to the local geometry of the aperture and inner wires. They can be measured in a laboratory, or computed with a 3D code, on a local meshing. Ecc and Hcc are the "short-circuit" electric and magnetic fields measured on the small aperture when it is closed with a metallic panel. These fields can be easily measured or computed with a 3D code. Consequently, the fact of closing the aperture allows to rigorously ignore the complexity of the internal volume in the resolution of the external problem. This example reveals also itself as a good demonstration of the capability of the theory to integrate the solution of local problems. These ones are treated for a long time, but are very rarely integrated in the complex context of the EM coupling on a system. Fig. 8 presents the derivatives of Ecc and Hcc fields on the caisson aperture, computed with the finite difference time domain (FDTD) 3D code ALICE, developed at ONERA. Fig. 9 depicts the comparison of a voltage obtained on a 50 Ohm load, by direct measurement and by introducing Ecc and Hcc fields computed at the external problem resolution step.

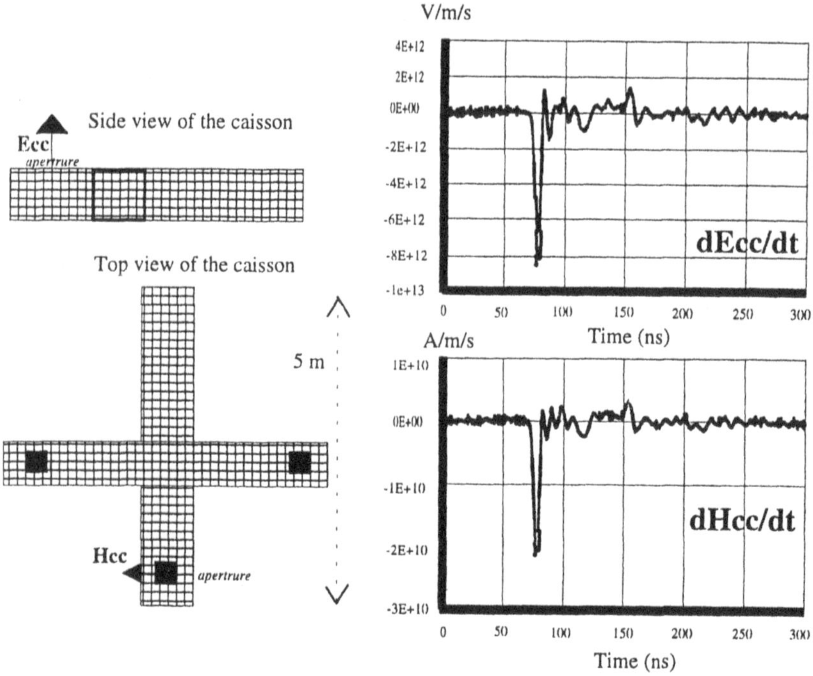

Figure 8: Derivative of "short-circuit" electric and magnetic fields computed with ONERA's ALICE FDTD code and measured on a caisson tested under CEG's SSR simulator.

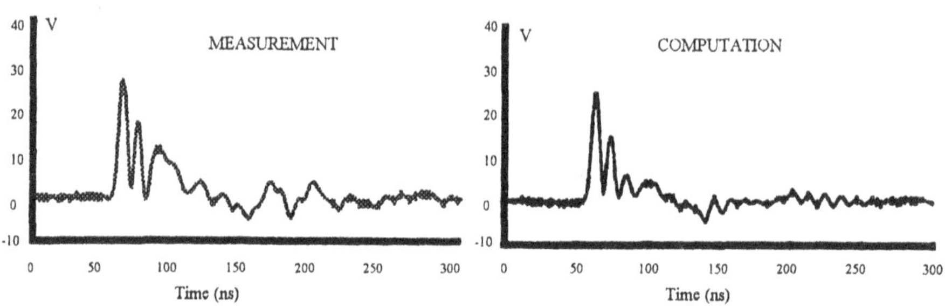

Figure 9: Comparison of a voltage measured and computed on 50 Ohm in the caisson placed under the CEG's SSR simulator.

5.3 Applying decomposition into elementary volumes

5.3.1 General case

We saw that the number of proper volumes could be limited to the decomposition into the external problem, then into the internal problem where wires are removed, and, finally, into the coupling on wires. Nevertheless, for each shielding level, each associated topological network has to be solved on its global geometry. Generally, these networks may still present a large size, and are characterized by a large amount of unknowns. Particularly, the wiring cannot be spread in independent networks as in the Good Shielding Approximation. Indeed, it is frequent that wires in a piece of cable harness run in other cable harnesses, in other volumes.

Consequently, one could think that the topological analysis of a problem has to be limited to the treatment of the only three steps previously mentioned, what could strongly restrict the interest of the topological decomposition. Fortunately, it is still possible to go on breaking down the problem in sub-problems. For this, the method cannot be based on an approximation. It deals with rigorously concatenating all the network information contained in a piece of network named a **"sub-network"**, in an equivalent junction associated to equivalent voltage generators coming from internal sources distributed on this sub-network. There is no precise rule for defining the limits of sub-networks, but it appears straightforward to delimit one sub-network in each elementary volume. Fig. 10 shows the principle of concatenation of a sub-network with the terminology used in ONERA's CRIPTE code. **"Internal junctions"** define the equivalent junction: they are junctions localized inside the concatenation limits (in dotted lines). **"Internal tubes"** relate two internal junctions and **"external tubes"** relate an internal junction to a non internal junction. In CRIPTE code, the output ports of the equivalent junction are computed at this level. The equivalent junction only characterizes the whole internal scattering inside the sub network and is fully independent of sources applied on it. So, the junction can be described with an S-parameter matrix calculated thanks to the individual S matrices of internal junctions and Γ matrices of internal and external tubes [*Parmantier*, 1991]. In a simple way, the equivalent vector generator can be seen as a generalisation of "Thevenin" equivalent generator for multiport system. Consequently, this voltage vector is equal to the open circuit voltage induced on the output port of the equivalent junction. Fig. 10 also shows how the resultant equivalent junction (defined on network A) can be used in other topological networks (for example in network B).

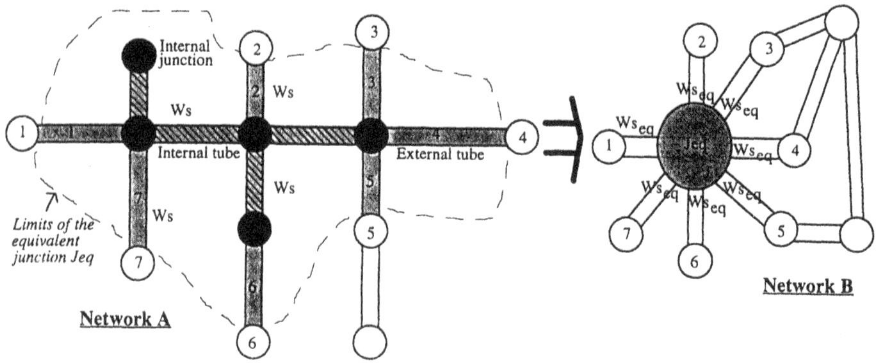

Figure 10: Schematic terminology of a sub-network (cf CRIPTE code terminology) and principle of the use of the equivalent junction and generators in different topological networks.

5.3.2 Example of cable network concatenation

The following example is an illustration of the sub-network concatenation utility for the treatment of large scale systems. It is taken from a work, performed by ONERA and Labinal companies, and supported by GEC Alsthom company, on the wiring of a subway train project, now in operation in Paris, called "Meteor". The subway train is made of four motor coaches placed in the centre and two slip coaches located at the extremities (see Fig. 11). In each coach, a wiring is installed on the under-frame. Perturbing sources are delivered by internal equipment. This wiring was interconnected from one coach to the neighbourhood coaches. The entire wiring network model for all the subway train was made of 158 tubes, 154 junctions and 1724 unknowns. The biggest junction had 39 ports and the biggest tube had 15 wires. The topological treatment dealt with defining six equivalent junctions limited to the six elementary volumes made by the coaches. In each volume type (motor or slip coaches), the same network could be duplicated because of the similitude of the wiring in each coach. Fig. 12 describes the final equivalent network dealing with the global subway train wiring. Each equivalent junction is related to the other by a simple junction Jc. Satellite junctions around equivalent junctions J1 and J2 are associated to the terminal loads on which an electrical signal had to be calculated. This junction could be used to study different types of connections of wires between two coaches. The important lesson to retain from such a study relies on the methodology: the resulting network is much smaller than the spread network one, and by this way, reveals itself as very well adapted to a modular approach. The final equivalent network contains only

246 unknowns (that is to say, 5 times less than the spread network), and is made of 27 tubes (with 27 Thevenin equivalent generators), and 28 junctions.

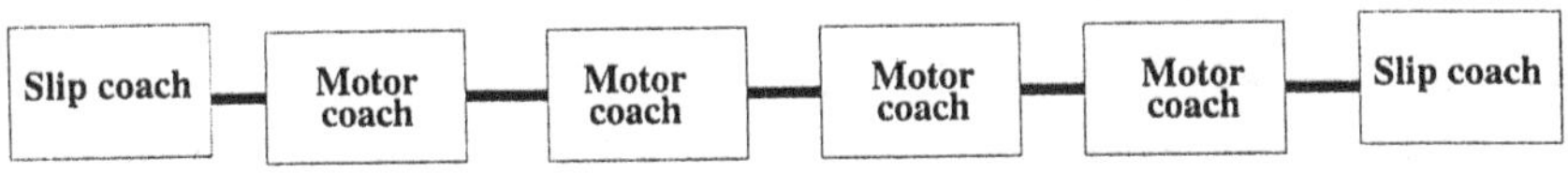

Figure 11: Schematic description of the Meteor subway train.

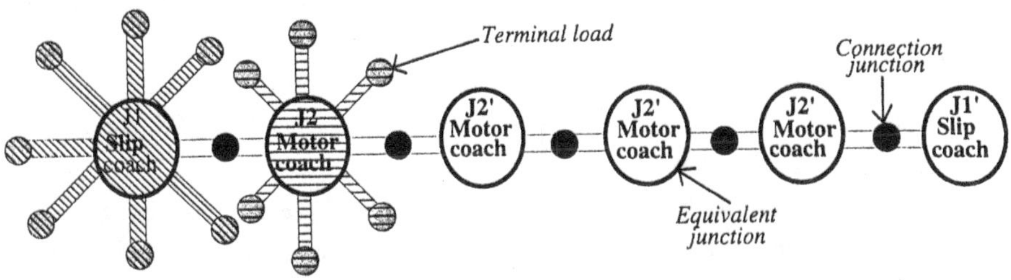

Figure 12: Resultant network for the under-frame wiring study after sub-network concatenation.

5.3.3 Concatenation with 3D codes

Let us clarify that the concatenation method is also available for networks dealing with physical volumes, different from wires (step 2). Precisely, junctions are like black boxes which summarize the scattering inside the volume. In this way they are similar to equivalent junctions presented before. Nevertheless, in this case, the characteristic of such networks is to contain a small number of tubes and junctions, but with large size matrices.

Until now, such methods have been applied to connect several numerical codes. This is the case for "Gemmacs" environment for which the external problem is treated by TGD, and the internal problem either by a Method of Moments or FDTD method [*Coffey et al.*, 1993]. Nowadays, the use of topological networks is being generalized at ONERA to perform such a numerical code hybridization. The network provides the connections between volumes, but also the coupling of unknowns inside the volumes (especially for asymptotic methods). Particularly, the properties of the scattering matrix of the network is used to reduce the data storage and the computation time and to facilitate the parallelization of calculations [*Michielsen*, 1993; *Ferrières and Michielsen*, 1994].

6 EMT in experimental works: application to EMPTAC aeroplane

6.1 Description of the EMPTAC

We previously insisted on the modular interest of EMT. Especially, we mentioned that measurements could be substituted to computations to determine the characteristics of tubes and junctions. Moreover, thanks to a real time analysis of results and their confrontation to measurements, this method can be very useful in experimental works, and can provide significant help for understanding EMC coupling paths inside the system. For the first time, in November 1993, ONERA and CEG, in collaboration with the Philipps Laboratory, had the opportunity to validate all the concepts and apply on an aeroplane, the EMPTAC (EM. Pulse Test Bed Aircraft), available at the Kirtland Air Force Base, in Albuquerque (New-Mexico) [*Parmantier et al.*, 1993, 1995] the methods developed during the last few years.

The EMPTAC is a 707 Boeing aeroplane. In the past fifteen years, it has been used to help on the design of specific aeroplane systems, in hardening programs, and was considered as a real laboratory to validate hardening technologies. With this objective, the original 707 aeroplane has been modified applying qualitative EMT rules. The aeroplane contains two volumes especially well shielded: the "lower forward shielded volume", and "the aft shielded volume". Each of these volumes is connected thanks to a specific electrical cable network. Out of both shielded volumes, the wiring is made of over shielded cables, maintaining, by this way, a homogeneous shielding level for the whole system (Fig. 13).

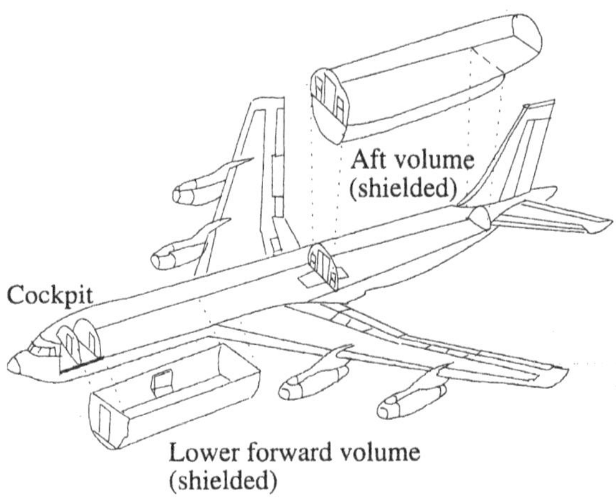

Figure 13: EMPTAC aircraft after modifications.

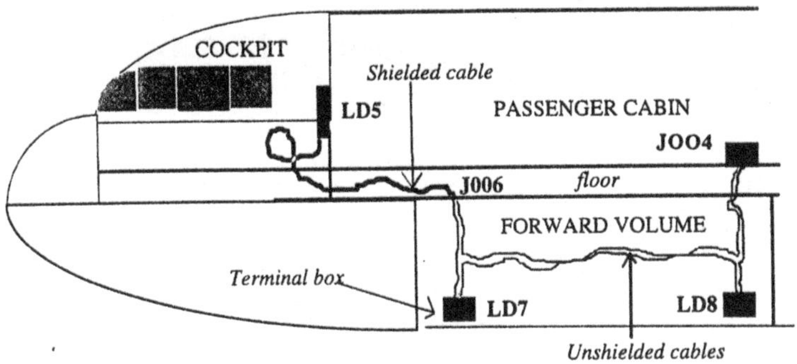

Figure 14: Topology of the problem under study and topological network.

During this experiment, the work was restricted to the coupling on a piece of wiring, running inside the cockpit and the forward shielded volume (Fig. 14): this wiring was called "network A". In the cockpit, the wiring under study was made of a 21 wire shielded cable running from the LD5 box, to the "forward shielded volume" with a separating connector JOO6 between them. At this point, inside the lower volume, the 21 wire cable became unshielded and ran along two paths, connecting LD7 and LD8 boxes. The LD7 box is also connected to JOO4 by means of unshielded wires.

The preparation of the experiment, held in France, suffered from the limited knowledge on network A: the exact running of the cables, their internal geometry were unknown. Only the load values (spread from 1Ω to 1MΩ) inside each black box were given in plans. Consequently, a pre-experiment was previously carried out in France by considering a similar network, containing only a few elementary wires, but presenting the same topological characteristics as Network A; that is to say, same running paths, same box location, same lengths. This test wiring was called "Network B". It was made of a 3 wire shielded cable put above a ground plane, corresponding to the cockpit volume, then becoming unshielded as in the forward volume. The connection between the two sides of the volume was made by means of 2 wire cable.

Fig. 15 presents the equivalent topological network associated to this problem. Each elementary volume has been represented with an equivalent junction. Each junction is itself the concatenation of the network existing in the associated volume. On the tube connecting the shielded cable equivalent junction, an equivalent Thevenin source has been placed to represent the coupling in this volume. The wiring inside the forward shielded volume has been designed, separating the terminal loads on which measurements were made. Finally, junction Jc makes simply the connection of junctions equivalent to the cockpit and the forward shielded volumes.

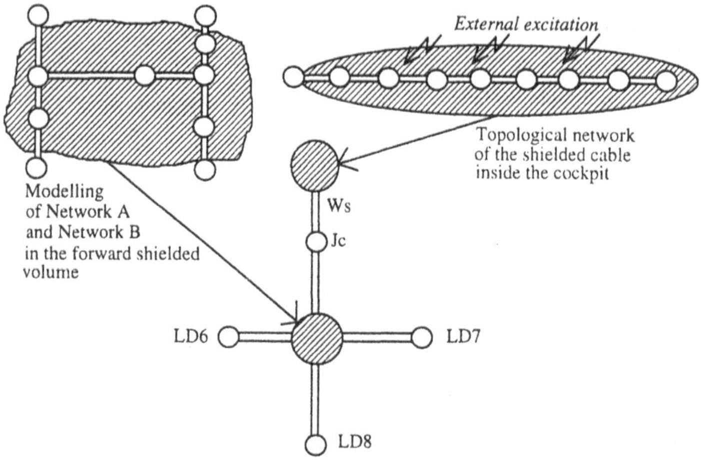

Figure 15: Principle of topological decompositioin for the study of the coupling in the cockpit and the forward shielded volume.

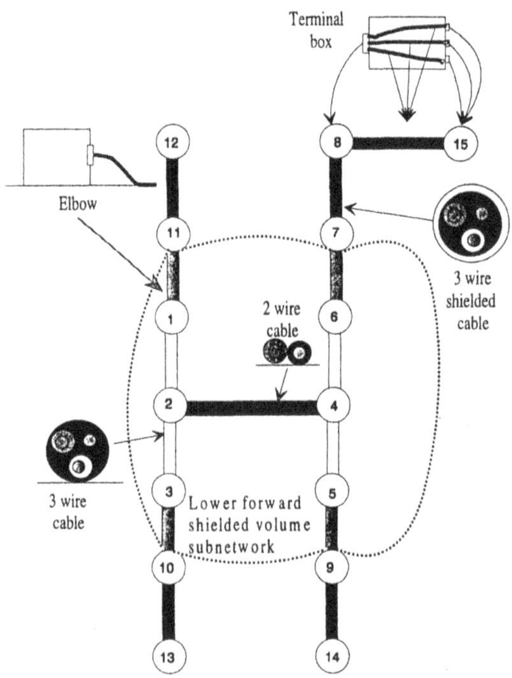

Figure 16: Topological model of network B on the metallic ground plane.

6.2 Test wiring pre-experiment study

The objective of the test wiring experiment was to constitute a large data bank on a wiring, representative of the real cable harness under study. For this, network B was tested after being installed on a metallic ground plane. First, the per unit length impedance and admittance of the elementary cables (that is to say, one, two and three wire cables, shielded and unshielded) were measured very precisely. The interest of taking into account the frequency dependence of these parameters was clearly demonstrated by making comparisons between direct S-parameter measurements and CRIPTE code numerical simulations: thus, a very good agreement was obtained, up to 500 MHz. In this canonical experiment, the "elbows" due to the junction between the cable and the box connector were modelled as an equivalent tube (Fig. 16), [*Besnier and Degauque*, 1995].

The immunity of the shielded cable was also tested in terms of transfer impedance Zt. The characteristics of the tube, taking Zt into account, were optimized by making comparisons between measurements and numerical simulations performed with CRIPTE code. The first important point in the model is that an injection current probe was used and considered as a transformer, inducing a tangential field in its centre. The second point is that the shield of the cable was considered as a real wire, as the other ones. Both models allowed to take into account all kinds of wire connections by simply changing some junction S-parameters in the topological network model. For example, the same model could be used to determine the response of the shielded cable, whatever the connections of its shield were.

6.3 In situ test wiring EM coupling analysis

The pre-experiment held in France allowed to constitute a data bank on network B elementary cables, and on the source generators produced by the current injector. All this data was used to feed the in-situ topological network models during the experiment performed on the aeroplane.

Indeed, the first phase of the in-situ experiment dealt with introducing Network B in the aeroplane and applying the same current injector source as the one applied in France. In fact, the topology inside the aeroplane was quite different from cables laying on a metallic plane (Fig. 17). During the experiment, it was impossible to reach such a detailed modelling as during the pre-experiment. Nevertheless, the cockpit shielded cable was simply decomposed with success into 10 tubes. Their lengths were determined thanks to reflectometry measurements. In the same way, the wiring inside the forward shielded volume was simply modelled by a network including characteristics of flattened and not flattened tubes, previously determined in France.

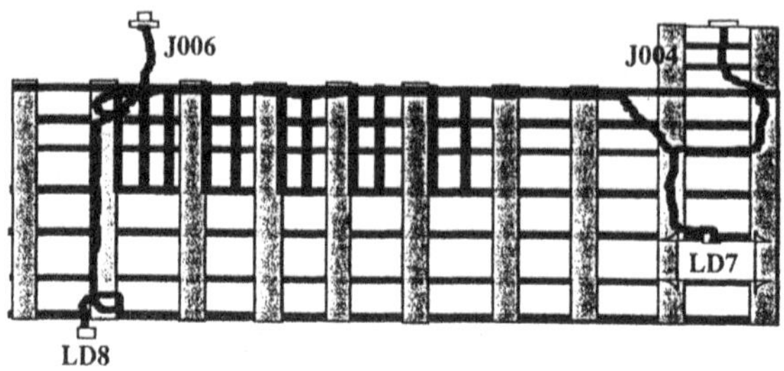

Figure 17: Scheme of the wiring topology in the lower shielded volume.

The next step of the experiment was to illuminate the whole aircraft with the "Ellipticus antenna". This antenna is an elliptic shape cable radiating a quasi plane wave. The principle of the simulator is quite similar to the concept of HPD in EMP, except the fact that this antenna works in the frequency domain, up to 1 GHz. Fig. 18 is a scheme of the experimental set-up.

The main problem for the numerical simulations was to determine the source terms coupling on each wire. The approximation that only the cockpit volume shielded cable was excited was made. In this frequency range, it was then supposed that the source could be reduced to a single generator. Its value was obtained by deriving an inverse problem. The current due to a 1 volt per meter generator was first calculated. The ratio, between the measured current and the current calculated at the same point, gave the generator.

The main interest of this generator lay in the fact that it was totally independent on the number of excited wires (the same generator was applied at the same location on all wires) and on the loads connected to wires. Fig. 19 presents an application when the shield of the cable is short-circuited or in open circuit at J006 point (see Fig. 17). Exactly the same topological network could be used to make the calculation, except the fact that S-parameters of the junction Jc were modified on the network depicted on Fig. 15. Both sets of curves are a comparison between a calculated and a measured voltage corresponding to the two kinds of shield connections.

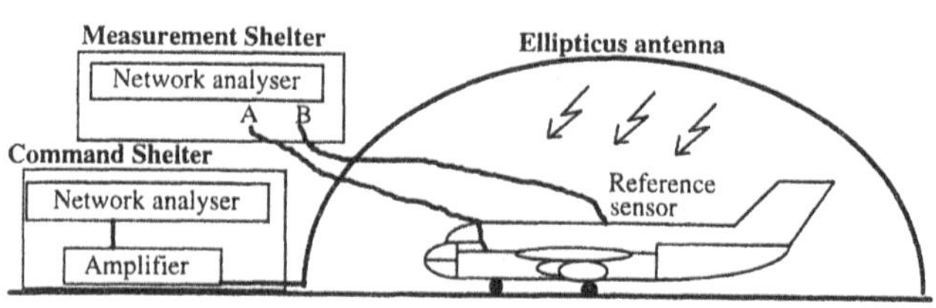

Figure 18: EMPTAC under the Ellipticus antenna.

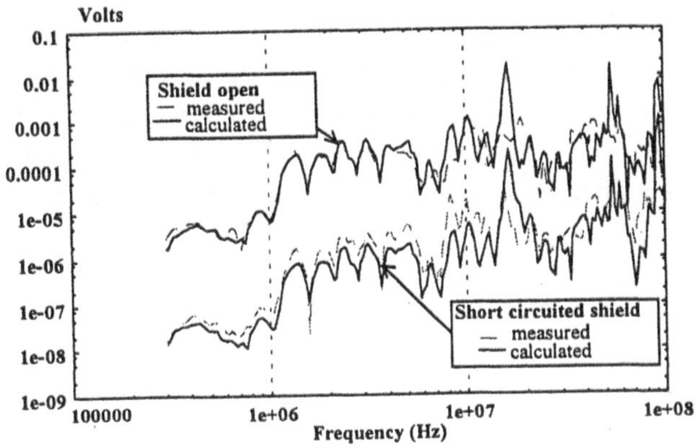

Figure 19: Calculated and measured voltage at LD8 load when the shield of the cockpit shielded cable is connected, or is not connected at JOO6 point.

6.4 EM coupling on network A

No electrical data concerning the geometry of cables constituting network A were available before the experiment. R, L, C, G parameters of the tubes had to be determined by estimating the geometry (radius of elementary wires, nature of their insulating dielectric, and respective position inside the harness). A 2D numerical code developed at ONERA, named LAPLACE, based on the resolution of Laplace equation by the Method of Moments was first used to calculate these parameters on a cross section of the tube. Then, to take into account the twisting between the wires, final tube models were obtained, by randomly connecting several lengths of the previously derived untwisted cable.

Because of lack of time during this experiment, only the coupling with the injection current probe could be performed. Fig. 20 gives a typical comparison between the measured and calculated voltage on a terminal load.

The injector was located in the cockpit shielded volume, and the characterization data file of the probe, already used on network B, could be used on network A. The discrepancies between the two curves are mainly due to the rough characterisation of the cable harness. Nevertheless, one must realize that these results come from a full predictive approach.

6.5 The teaching of EMPTAC experiment: complex wiring test procedure

To our knowledge, this experiment was the first full application of quantitative EM

Topology concepts on a complex system. It showed the interest of using a test wiring to characterize the internal coupling in terms of electric fields distributed along the running of the actual cables. The results, obtained on the cable harness, also demonstrated that a predictive approach, based on the use of a numerical code, could be reliable on a complex wiring system. Particularly, the modularity of the method was pointed out by using data, which had been previously determined in France, before the experiment.

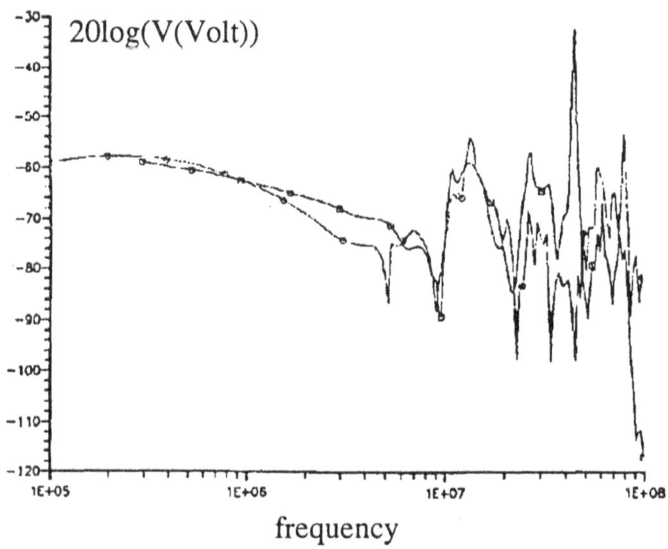

Figure 20: Comparison of induced voltages on network A (LD7 box): O Calculated; X Measured.

In June 1995, a new campaign in collaboration between ONERA and CEG has been held on the same aircraft with the objective of testing the numerical tool capabilities from 100 MHz up to 1 GHz.

7 Conclusion

Nowadays, Electromagnetic Topology can be considered as a helpful method to help to the design and analysis of EMC in large scale systems. Whereas before 1990, this method existed at a state of theory, many adaptations and validations of the concepts have been developed for five years. Based on a network formalism, numerical tools have been developed (cf. ONERA's CRIPTE code). In 1995, two domains of applications can be distinguished for EMC analysis on large scale systems. The first domain is called "wiring topology" and deals with a frequency spectrum from DC to

500 MHz, spectrum where coupling on cables remain predominant in EMC. From 1995, numerous validations have been performed in this frequency range and the method begins to be applied in industrial companies. As if fundamental concepts are now well defined, several progresses could appear in future years:

a - the modelling of non linearities of equipment input impedance,

b - the application of matricial norms for the calculation of EM susceptibility margins,

c - the modelling of cable radiation.

The second domain deals with higher frequencies, domain in which cable are less critical for EMC compared to wiring topology. This domain is still at a development stage. Because of the complex nature of EMC phenomena, more sophisticated and heavy technical methods than the one used in wiring topology have to be used. Nevertheless, the interest of EM Topology would lead in the future to the treatment of EM coupling in large scale problems compared to the wavelength under study. Nowadays, the applications in this domain lead to an hybridization of 3D numerical technical methods and provide promising results.

References

Agrawal, A.K., K.M. Lee, L.D. Scott, H.M. Fowless [1979], "Experimental Characterization of Multiconductor Transmission Lines in the Frequency Domain"; *IEEE Transactions on EMC*, 21, 1, 20-27.

Baum, C.E. [1982], "Sublayer Sets and Relative Shielding Order in Electromagnetic Topology". *Electromagnetics*, 2, 335-354.

Baum, C.E. [1989], "The Theory of Electromagnetic Interference Control", *Interaction Notes, Note 478*, and *Modern Radio Science*, Oxford University Press, 87-101.

Baum, C.E.,T.K. Liu, F.M. Tesche [1988], "On the Analysis of Multiconductor Transmission Line Networks", *Interaction Notes, Note 461*.

Besnier, P. [1993], "Etude des couplages électromagnétiques sur des réseaux de lignes de transmission non uniformes à l'aide d'une approche topologique", *Lille University Thesis Report*, France, February (in French).

Besnier, P., P. Degauque [1995], "Electromagnetic Topology: Investigation of Non uniform Transmission Lines Networks", *IEEE Trans. EMC*, 37, 2, 227-233.

Clark, L.H., C.E. Baum [1988], "Realization of Sublayer Relative Shielding Order in Eletromagnetic Topology", *Interaction Notes 471*.

Coffey, E.L., N.W. Coffey, D. L. Kadlec [1993], "Gemacs Source Book, Version 5.0", *Advanced Electromagnetics*.

Degauque, P., J. Hamelin [1990], *Electromagnetic Compatibility*, Oxford University Press

Ferrières, X., B.L. Michielsen [1994], "Resolution efficace des équations de lignes de transmission pour réseaux complexes (in French)", *Proceedings of the French EMC Symposium, Toulouse*, 283-287.

Gobin, V., F. Issac, J.P. Aparicio, J.P. Parmantier, I. Junqua, A. Madore, D. Gautier [1992], "Validation de la Topologie Electromagnétique sur une structure complexe (in French)", *Proceedings of the French EMC Symposium*, Lyon-Ecully, 105-110.

Karlson, T. [1978], "The Topological Concept of Generalized Shields". *Interaction Notes*, 350.

Lee, K.S.H. [1980], "A Complete Concatenation of Technology for the EMP-Principle, Technique and Reference Data", *Interaction Notes.*

Michielsen, B.M. [1993], "Path-sum Formulas in Transmission Line Network Analysis", *Proceedings of EMC Symposium, Zurich*, 239-244

Noss, R. [1983], "Planarity Criteria in Electromagnetic Topology". *Interaction Notes*, 434.

Parmantier, J.P., [1991], "Approche topologique pour l'étude des couplages électromagnétiques". Lille University thesis report. Translation in English available at ESA.

Parmantier, J.P. [1993], "Recent Developments in Electromagnetic Topology", *Transactions of the 3rd ESA workshop. Pisa (Italy)*. 3-10.

Parmantier, J.P., J.P. Aparicio [1991], "Electromagnetic Topology: Coupling of two Wires Through an Aperture", *Proceedings of EMC Symposium, Zurich*, 595-600.

Parmantier, J.P., V. Gobin, F. Issac, I. Junqua, Y. Daudy, J. M. Lagarde [1993], "An Application of Electromagnetic Topology on the Test-Bed Aircraft, EMPTAC". *Interaction Notes*, 506.

Parmantier, J.P., V. Gobin, F. Issac, I. Junqua, Y. Daudy, J.M. Lagarde [1995], "Analysis of EM Coupling on Large Structures using EM Topological Concepts: application to the EMPTAC aircraft", *Proceedings of the Inter. Conf. on Electromagnetics in Advanced Applications* (ICEEA 95), Torino, 81-84.

Parmantier, J.P., G. Labaune, J. C. Alliot, P. Degauque [1990], "Electromagnetic Topology on Complex Systems: Topological Approach". *Interaction Notes*, 488, and *La Recherche Aérospatiale*, 5, 57-70.

Parmantier, J.P., G. Labaune, J. C. Alliot, P. Degauque [1992], "Electromagnetic Topology: Junction Characterisation Methods". *Interaction Notes*, 489.

Vance, E. [1978], *Coupling to Shielded Cables*, Wiley - Interscience Publications, 1978.

Vétri, J.L., G.I. Costache [1991], "An Electromagnetic Interaction Modelling". *IEEE Trans. on EMC*, 33, 3, 241-251.

Yang, F.C., C.E.Baum [1986], "Electromagnetic Topology; Measurements and Norms of Scattering Parameters Subshields". *Electromagnetics*, 6, 47-59.

IMPACT OF NUMERICAL METHODS ON PROPAGATION MODELLING

Kenneth H. Craig

1 Introduction

Recent years have seen a growth in the development of numerical tools for radiowave propagation. This chapter considers some of the reasons for this development, the benefits of simulation in radiowave propagation, and gives specific examples. Numerical models are of course being applied in many electromagnetic wave propagation problems. However the scope of this paper will be limited to tropospheric applications dominated by refraction, diffraction and scatter mechanisms. This includes several practical problems in radio system design; for example mobile and personal communications, microwave radar design and operation, and the design of millimetre wave systems in the urban environment.

The potential benefit of a numerical model that can simulate and predict the performance of a radio or radar system is clear. For example, while the design of a mobile radio network can be planned using semi-empirical or statistical methods followed by in-situ site surveys to ensure that a complete coverage can be provided, a high fidelity numerical model would reduce the need for expensive, manpower intensive, measurements. Another case is predicting the performance of radars in the presence of atmospheric effects (such as ducting) that can distort the radar coverage. In-situ measurements of radar performance may be impossible or impractical, and numerical models may be the only viable method of providing the required information. Indeed numerical models can go even further by providing a *forecast* of radar performance some time ahead, if a weather forecast of sufficient fidelity is available.

In principle, a solution of any electromagnetic wave problem is provided by

Maxwell's equations. A model of this type is *deterministic* in the sense that if the propagation environment is characterised *exactly*, the solution would also be an exact description of the propagation conditions for the given path at the given time. In reality approximations and simplifications are required, and no practical solution of Maxwell's equations (that is, one that is analytically solvable or is numerically tractable) will be applicable to all propagation problems. Different propagation mechanisms require different approximations, and several methodologies have been developed to deal with the different application areas.

Deterministic models are particularly appropriate when a prediction (or even a forecast) of the performance of a specific system is required at a specific location and time. However, the accuracy of a prediction is critically dependent of the quality (and perhaps quantity) of real-time environmental data available; the (lack of) availability of such data is often a more serious constraint on the prediction accuracy than the accuracy of the propagation model itself.

It is in their approach to fidelity that the capabilities of numerical and analytical approaches differ. The information provided by an analytical (or theoretical) model will indeed be exact, but only in the rather rare cases where the assumptions of the theory are satisfied; since most theoretical models impose rather severe constraints on the problem domain, "rigorous" analytical methods may be unable to capture the complexity required in real life. Numerical models are generally able to handle a much wider range of input variables and are therefore more likely to represent faithfully the complex physics even if, in some cases, the numerical model may make mathematical approximations.

Empirical or statistical predictions may be necessary when no direct measurement of the environment is available, or is very imprecise. These can be based on historical data concerning the environment, or on measurements of the performance of similar systems in similar propagation conditions. This type of model has in the main dominated the planning of radiocommunications systems, particularly in the civilian sector [*ITU-R*, 1994].

There are several reasons why numerical methods are now becoming increasingly important:

- The ready availability of powerful desk-top computers has been a major stimulus. Although numerical methods were applied to propagation problems in the earliest days of the computer (during the Second World War), the practical applications were limited because of the costs involved. Application to civil problems had to await the wide availabilty of powerful, inexpensive, computers. Not only are the speed of these now such that quite complex models can be run in an acceptable time, but interactive user interfaces have enabled these to move out of the research laboratory onto the desks of the non-specialist user.

- In recent years the quality and availability of environmental data has greatly increased. This is partly due to improvements in the quality and capability of environmental sensors (for example satellite remote sensing) and partly due to the enormous increase in the amount and resolution of data available (which in turn is made possible by the ready availability of high capacity data storage that is available at low cost).
- New radiocommunication services at higher frequencies require better prediction tools than previously. For example, statistical methods are of limited value at millimetric frequencies for the design of urban point-to-point and point-to-area links. The siting of individual buildings or hilly terrain can produce strong shielding or provide a source of high reflected field levels which could affect the communications channel. In this case, a *site specific* deterministic model is required.

The rest of this chapter describes two applications where numerical models have made a recent impact. The first is the modelling of atmospheric refractive bending, and was one of the earliest applications of numerical models to tropospheric radiowave propagation, although the last ten years have seen the development of new numerical techniques. The second is the application of an old technique (ray tracing) to urban propagation; this has become viable because of the availability of site specific building databases and powerful computing and graphics facilities.

The applications chosen are those most familiar to the author. The discussion is not intended as an exhaustive review, either of the history of the subject, or of existing models. Comparisons will be made between different numerical techniques in the hope that the reader can get a feel for current research and the reasons for choosing one method rather than another. This is not a mathematical exposition of the numerical methods themselves, although references to the literature are given where appropriate.

2 Numerical methods for refractive effects modelling

The first application of numerical methods to tropospheric radiowave propagation dates from the earliest days of numerical computing engines, and the impetus was the understanding of radar propagation during the Second World War. The methods developed then (ray tracing and waveguide mode theory) have continued in use until the present day, although in the last ten years parabolic equation methods have begun to replace these in practical propagation tools.

Much of the impetus for the continuing development of these techniques is still the radar coverage application. Shipborne radars are affected by the marine evaporation duct, and at microwave frequencies the effects can be quite strong and cannot be ignored. However, the advent of desk-top systems has also led to applications in civil communication systems and in system design. Also the educational value of these simulation tools in visualising the effects of the atmosphere on radiowave propagation should not be underestimated.

2.1 Early history

Many of the great names of radio science contributed to the early history of the theory of tropospheric wave propagation, including Poincaré, Fock, Van der Pol, Bremmer, Eckersley, Watson and Millington. A review of refraction theories of radiowave propagation in the troposphere by *Arora* and *Wait* [1978] describes this early history. The earliest work was analytical and aimed principally at the problem of diffraction around the spherical Earth with no atmosphere. However, the Second World War brought the need to extend the work to the case of an atmosphere, and atmospheric refractive index profiles that were too complex to solve analytically.

At about this time it was realised that the combined effects of spherical Earth diffraction and the atmosphere could be modelled by introducing a "modified" refractive index that allowed the problem to be solved over a "flattened Earth". This led to the development of mode theory [*Booker* and *Walkinshaw*, 1946] which was the basis for most numerical models of long distance (beyond-the-horizon) tropospheric propagation until the 1980s. Because mode theory imposed fairly strong constraints on the form of the refractive index profiles that could be modelled, optical ray tracing methods were also used.

The earliest application of computing machinery to the tropospheric propagation problem was reported by *Hartree, Michel* and *Nicholson* [1946]. Initially a differential analyser was used to evaluate ray paths through a stratified medium. However, difficulties in dealing with the (now well-known) problems of geometrical optics near caustics in ducting atmospheres focused the later work on solutions of the "full-wave" normal mode equations. The differential analyser was used to find the modal eigenvalues, and to evaluate the eigenfunctions. After this early work, both ray and mode methods were pursued by various workers.

2.2 Ray methods

Naïve ray tracing consists of simply "shooting" rays from the transmitter in a vertical plane with equally spaced ray launch angles. An example is shown in Figure 1. The rays follow paths determined by the vertical refractive index gradients at each range; in this case there is a strong trapping layer at a height of 80m. Note that the information obtained is qualitative. Only the ray paths are traced; the field amplitudes are not calculated. However, the ray paths do give an indication of where the radiated energy can propagate to, and the density of rays gives some indication of relative field strength. This range-height type of display is known as a vertical coverage diagram, and is very useful in assessing the performance of radar systems.

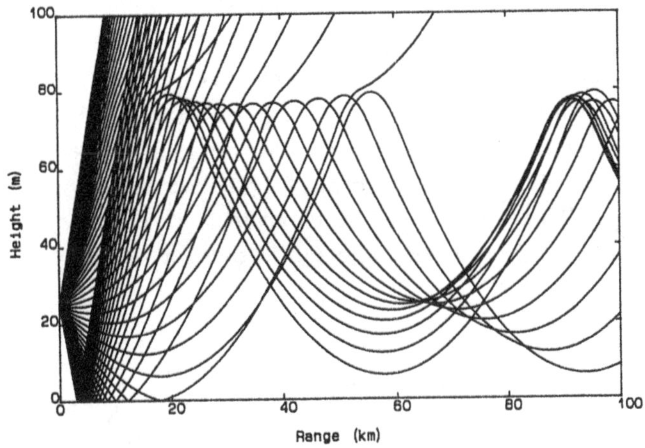

Figure 1: Ray tracing in an atmosphere with a trapping layer at 80m.

Quantitative results need a calculation of the ray amplitudes. This can be done either by explicitly calculating the cross-section of a pencil of rays or by integrating the transport equations. In both cases a careful *a priori* separation of rays into classes (for example, those that have undergone one layer reflection from those that have undergone two layer reflections) and the coherent addition of rays in different classes is required. This can be difficult to automate. The method is only applicable in those regions reached by the rays. In shadow regions (for example, above the trapping layer of Figure 1), and near caustics and cusps in the ray trajectories, ray tracing fails. Although the method can be extended to circumvent the limitations, these extensions have not found wide applications in practical tools.

Despite the limitations, ray methods have found ready application in predicting the effects of refractive layers on the performance of line-of-sight communications systems [see for example, *Shkarofsky and Nickerson*, 1982; *Webster*, 1982; *Craig and Kennedy*, 1987]. This application essentially requires a point-to-point calculation (i.e. transmitter to receiver), and the ray paths are obtained by an iterative method of "homing in" on the receiver location;

the separation into ray classes is simplified because the topology of the ray paths can be calculated in advance. Information on the effects of multipath propagation (such as ray angles-of-arrival, path delays and amplitudes) can be obtained; this has proved useful in the design of antennas and adaptive equalisers for digital systems.

Ray methods have also been applied to the problem of obtaining quantitative vertical coverage diagrams for radar systems [*Herring*, 1985]. However the main interest in coverage prediction for surveillance radars lies at long (beyond-the-horizon) ranges, and at these ranges it becomes increasingly difficult to maintain sufficient accuracy to keep track of the different ray classes contributing to the overall field. For these applications, mode theory, or a combination of mode theory and ray methods, found favour.

In the troposphere, vertical refractive index gradients are several orders of magnitude greater than horizontal gradients. It is therefore normally adequate when considering refractive effects to solve (a) a scalar wave equation since polarisation states decouple, and (b) a two-dimensional (range–height) problem since cross-path refractive bending is negligible. Three-dimensional, vector ray tracing is necessary in ionospheric propagation problems and has been addressed in the troposphere [*Shkarofsky* and *Nickerson*, 1982].

2.3 Mode methods

In mode theory, the transmission medium is modelled by analogy with a dielectric-filled waveguide, and by imposing suitable boundary conditions, Maxwell's equations can be solved for the propagation modes present. A good general text on mode methods is *Budden* [1961]. Mode methods can be purely numerical, or can be based on the numerical evaluation of analytical solutions for simplified refractive index profiles.

The earliest applications of mode theory were to the problem of spherical Earth diffraction and to propagation in shallow surface ducting layers (such as the marine evaporation duct). In these application, the number of modes that contribute significantly to the received field is limited, and often the refractive index profiles can be approximated by simple functions, making the problem tractable.

However, the atmospheric "waveguide" can be large compared to a wavelength at microwave and millimetre wave frequencies, particularly for propagation in the presence of an elevated tropospheric layer. This whispering gallery type of propagation [*Wait* and *Spies*, 1969] can give rise to the propagation of several hundred significant modes. The problem is intractable unless severe restrictions are imposed on the form of the refractive index structure. For a similar reason, mode theories are not practical for the prediction of field levels at short ranges (for example in the calculation of the interference lobing that is the principal effect of multipath at line-of-sight ranges). This is because the number of modes contributing to the overall field increases enormously as the range approaches the radio horizon.

Most mode models assume that the refracting layers are horizontally stratified; in other words the vertical refractive index profile is range-independent. While this is often a good approximation, it is not always true, and the errors can be large, particularly for predicting the effects of elevated layers on airborne radar coverage. Attempts have been made to include horizontal variability. One approach [*Cho* and *Wait*, 1978] models the atmosphere as a series of one-dimensional sections. The modes are calculated for each section as normal, and the coupling between the sections is determined by the "mode mixing" at the boundaries between the sections. These approaches are interesting, but very expensive computationally, and are not used in practical prediction tools.

Mode theory calculations being inherently point-to-point and numerically time-consuming, the calculation of a two-dimensional radar coverage diagram is prohibitively expensive. Despite the limitations of both ray and mode methods, these were very successfully incorporated into the first propagation prediction tool that was intended to satisfy the operational needs of users who were not propagation or numerical experts. This was the Integrated Refractive Effects Prediction System (IREPS) [*Baumgartner et al*, 1983], developed for the US Navy. IREPS uses a ray trace in the line-of-sight interference region, and a single mode diffraction model for the diffraction region under standard propagation conditions. Ducting conditions are accounted for using "pre-calculated" mode theory results by matching the actual atmospheric profiles to a set of "template" profiles. Because of the single propagation mode assumption, the effects of elevated ducts could not be quantified.

2.4 Parabolic equation methods

In recent years, an alternative, purely numerical, approach to the modelling of refractive conditions has emerged. This is the parabolic equation (PE) method. The name relates to the fact that the scalar wave equation

$$\nabla^2\psi + k^2n^2\psi = 0 \tag{1}$$

(where $k = 2\pi/\lambda$ is the wavenumber in vacuum, and n is the refractive index) can be reduced to a parabolic differential equation

$$\frac{\partial^2 u}{\partial z^2} + 2ik\frac{\partial u}{\partial x} + k^2(n^2(x,z) + 2z/a - 1)u = 0 \tag{2}$$

in two dimensions (range x and height z) for the "amplitude" function $u(x,z)$, obtained by removing the fast phase variation in x from ψ. a is the Earth's radius of curvature.

This reduction is valid if the refractive index changes relatively slowly with range, and is always satisfied for realistic refractive index structures in the atmosphere. The parabolic approximation is (a) paraxial and (b) represents forward scatter only, but these are not severe constraints for tropospheric refractive effects applications. As in the case of ray tracing, scalar, two-dimensional PE models are adequate for many tropospheric propagation problems, and the discussion will be restricted to these cases.

The PE was first introduced into radiowave propagation by *Leontovitch* and *Fock* [1946] for modelling spherical Earth diffraction. However, the lack of computing power at the time meant that the PE method was overshadowed by the semi-analytical mode methods. On the other hand, PE methods were developed by the underwater acoustics community for sonar propagation. The recent resurgence of interest in the PE for tropospheric propagation draws largely on the expertise built up in this field.

The PE of eqn (2) is actually a rather special (and simple) form of a more general equation:

$$\frac{\partial u(x,z)}{\partial x} = ik(Q(x,z) - 1)u(x,z) \tag{3}$$

where $Q(x,z)$ is an operator given by

$$Q(x,z) = \sqrt{1 + (n^2 + 2z/a - 1) + \frac{1}{k^2}\frac{\partial^2}{\partial z^2}} \tag{4}$$

A good exposition of the derivation and the different forms of the PE is given by *Lee and McDaniel* [1987].

There are two numerical approaches to solving the PE. The Fourier "split-step" algorithm [*Hardin* and *Tappert*, 1973] was the first to be applied to radiowave propagation [*Ko et al*, 1984], and has proved to be the most popular because of its relative efficiency and ease of implementation [*Craig* 1988; *Dockery*, 1988; *Craig* and *Levy*, 1991a; *Kuttler* and *Dockery*, 1991; *Hitney*, 1992]. The finite difference method was first introduced for refractive effects modelling [*Fournier*, 1989], but has found its main application in PE models that include the effects of terrain diffraction as well as refractive bending [*Levy*, 1990].

Both the split-step and finite difference methods integrate the PE with respect to range x by means of a "marching" technique on a discretised range-height (x-z) grid; this advances the amplitude $u(x,z)$ from range x to range $x+\Delta x$. By iteratively applying the algorithm, a series of vertical "stripes" are produced, beginning at some initial range x_i and ending at some final range x_f. The number of steps required depends on the step size Δx, which has to be small enough to satisfy some accuracy criterion for the solution; generally more accurate solutions require smaller Δx. In addition to the stepping algorithm, it is necessary to specify the vertical step size Δz and the boundary conditions to be satisfied at the top and bottom of the domain of interest.

The split-step and finite difference solutions of the PE are purely numerical methods and can deal with refractive index structures that are specified independently at each range-height point. No *a priori* assumptions about the profiles need be made. This is a major advantage for the new generation of propagation tools where *forecasting* relies on an accurate prediction of the state of the propagation medium; this is being achieved by means of numerical weather models, which predict meteorological parameters on a three-dimensional grid. Improvements in propagation forecasting are now closely tied to developments of numerical methods and data assimilation in this field.

Another feature of the PE method is that it deals in a uniform way with all ranges from the line-of-sight region out to the transhorizon region. The PE is therefore much more flexible than either ray or mode methods. Although it is often said that analytical methods give more "insight" into a physical problem than numerical methods, this is only true if the problem is sufficiently simple in the first place. The PE has given new insights into real propagation problems that could not be handled by previous methods. Figure 2 shows the PE coverage diagram corresponding to Figure 1 for a frequency of 9.5GHz. Note the quantitative treatment of the interference lobing, the leakage from the top of the duct, and in the region of the ray caustics.

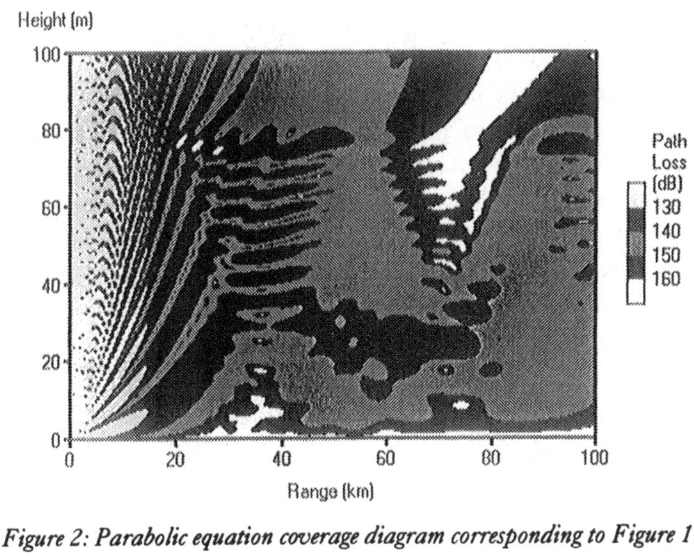

Figure 2: Parabolic equation coverage diagram corresponding to Figure 1.

One of the differences between split-step and finite difference solutions is the way that the surface boundary condition is implemented. All models use the Leontovich impedance boundary condition in one form or another. Early split-step methods implemented this via an optical image approach, but recently a more rigorous approach in terms of a discrete mixed transform (Kuttler and Dockery, personal communication, 1995) has been implemented. The finite difference scheme applies the Leontovich boundary condition explicitly via a finite difference implementation of the field gradient at the surface.

At the upper boundary, a numerical approximation to the Sommerfeld radiation condition is required. All PE models assume that above some finite height, higher than the domain of interest, there are no refractive index structures that could deflect energy back down into this domain. To prevent spurious reflections from the upper boundary itself, most PE models have applied an "absorber" or "filter" in a buffer region above the domain of interest that removes up-going radiation before it reaches the top of the integration grid. This method is fairly well-established, but does involve unnecessary computation in the absorber region which may need to be 50-100% of the size of the domain of interest in order to reduce reflections to an acceptable level. In recent years, non-reflecting boundary conditions have been applied to this problem. A promising approach is the use of non-local boundary conditions [*Dalrymple and Martin*, 1992] which have been applied to the tropospheric propagation problem [*Marcus*, 1991; *Levy*, 1994].

The first generation of practical PE models were based on the pure split-step or finite difference methods. However, the run times of both "pure" methods can be high for "large" problems. The split-step method is more efficient because of its better convergence properties. (Here we are discussing errors due to the numerical implementation, as opposed to approximations in the PE reduction itself). The operator Q is implemented via a Fourier transform, and does not require numerical approximation; the split-step solution is also *exact* in vacuum within the PE assumptions and in this case there is no upper bound on the range step size Δx. On the other hand, the finite difference scheme approximates the range and height derivatives and is not exact even in vacuum; the convergence can be slow, and requires smaller step sizes for higher frequencies and propagation angles. The computation times for the split-step T_{ss} and the finite difference T_{fd} schemes are:

$$T_{ss} \propto k\theta_{max} x_{max} z_{max} \tag{5}$$

and

$$T_{fd} \propto k^2 \theta^3_{max} x_{max} z_{max} \tag{6}$$

where k is the wavenumber, θ_{max} is the maximum propagation angle to be represented,

x_{max} and z_{max} are the maximum range and height of the domain of interest, respectively. Although PC-based practical prediction tools based on these methods are available [*Craig* and *Levy*, 1991b], run times can be high for some problems.

Two of the main sources of large run times in a "pure" PE approach are actually avoidable for physical reasons:

a) high angle radiation (greater than a few degrees to the horizontal) is not affected by the weak atmospheric refractive effects that occur in practice; thus it is not necessary to use a rigorous full-wave solution like the PE to model high angle energy. The main reason for including high angle radiation is simply to represent antenna pattern effects in radar coverage diagrams (where beamwidths may be large, even omnidirectional).

b) Refracting layers strong enough to trap energy are very unlikely to occur above a few kilometres in height. For a surface-based antenna a pure PE simulation from the ground up to a height of 10km is unnecessarily extravagant.

Significant speed-ups have recently been achieved by the introduction of hybrid numerical methods. The range-height domain is divided into several subdomains, and a pure PE only used in that region (low-angle, low-altitude) where it is essential. More efficient methods are used in the other subdomains. A good example of this type of model is the Radio Physical Optics (RPO) model [*Hitney*, 1992] which uses an optical ray trace for high angle radiation, and an "extended optics" ray method for the high altitude region. An alternative treatment of the high altitude region is the "horizontal PE" [*Levy*, 1995] that extends the field vertically upwards by means of a PE in which the range and height variables are interchanged.

Finally, PE methods have emerged as a very convenient approach to terrain diffraction modelling. The earliest applications were based on the finite difference solution [*Levy*, 1990; *Marcus*, 1992], but more recently, split-step solutions have been introduced [*McArthur*, 1992; *Barrios*, 1994]. The PE is at present the only method that can deal with terrain and refractive effects in a unified way, and much work is currently devoted to developing a truly user-friendly, efficient, terrain-refractive-effects simulation/prediction tool.

Figure 3 shows the coverage diagram for an omni-directional 3GHz emitter in the presence of terrain and a ducting atmosphere. Despite the large range–height domain, the calculation took about a minute on a desk-top computer using a hybrid vertical/horizontal PE and ray trace method.

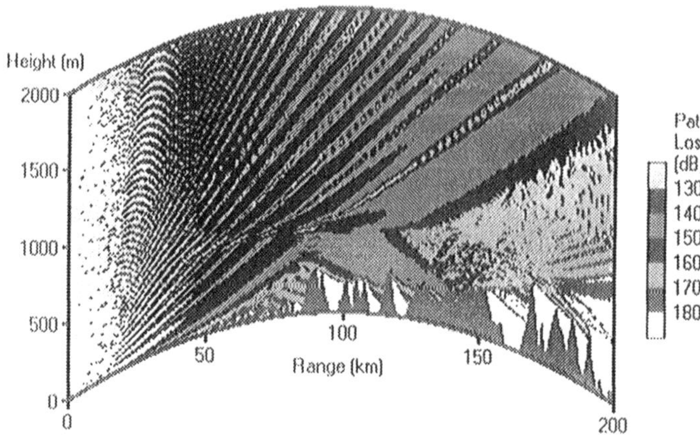

Figure 3: Hybrid parabolic equation modelling of refraction and terrain diffraction.

3 Site specific propagation models

In many countries, civil analogue and digital mobile radio and Personal Communications Systems (PCS) have developed rapidly in the last few years. These services have given rise to the requirement for planning tools that can predict service area and interference problems (particularly important as the density of systems increases). Operating at higher UHF frequencies (for example 900 MHz and 1800 MHz), the semi-empirical urban clutter models used at lower frequencies are less appropriate. Instead, coverage becomes much more dependent on the details of the ground cover (particularly buildings in urban areas) between transmitter and receiver, and "site specific" models based on building databases are now being developed to account for this. These models can also be applied at higher frequencies: satellite-mobile coverage prediction and the siting of small aperture terminals at microwave frequencies, and point-to-point and point-to-area services at microwave and millimetric frequencies.

Any numerical model for these applications must account for several propagation mechanisms. The mechanisms will depend on application and frequency, but will include at least building scatter and (at lower frequencies) diffraction. Because the scattering environment is very complex, ray tracing has been the preferred approach, rather than a direct solution of the wave equation. There are two distinct approaches to ray tracing which will be discussed below. This is followed by a consideration of the propagation mechanisms that need to be included, other approaches to modelling, and the building databases necessary for propagation modelling.

3.1 Ray tracing

In the last few years, several groups have been developing tools for site specific urban propagation based on ray methods. Most of these have been aimed at the mobile communications application in the UHF band, although they have also been used for broadcast, fixed link and indoor propagation. There are now commercial packages available.

Ray tracing as used for urban propagation is rather different from the modelling of refractive effects in the atmosphere. The relatively short propagation paths mean that refraction can be ignored. The rays are then straight lines through the atmosphere, but undergo many possible interactions with the buildings and terrain, for example scatter from building surfaces, or diffraction round building edges.

Ray tracing consists of finding all the ray paths that connect a transmitter and a receiver location and which take into account all the significant propagation mechanisms (such as reflection and diffraction). As an example, Figure 4 shows the *reflected* ray paths connecting a transmitter and receiver in a small section of a building database. The level of building detail shown is typical of current databases.

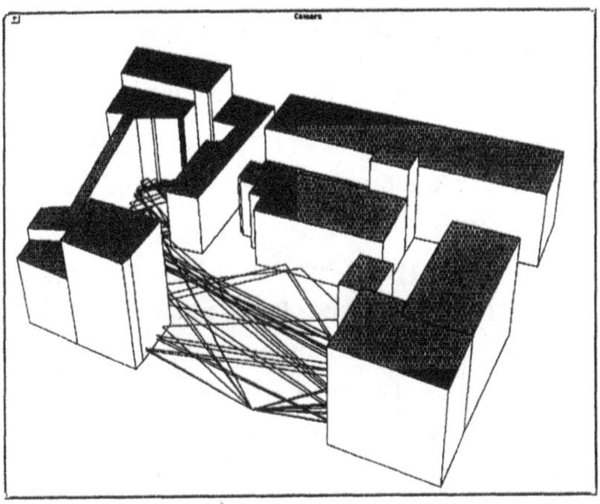

Figure 4: Reflected ray paths in a small section of an urban building database.

The received signal is obtained by coherently adding the contribution of each ray. The amplitude and phase of each ray is obtained by considering (a) the free-space path loss due to the end-to-end path length of the ray; (b) the amplitude reduction and phase shift caused by each interaction of the ray with a building or the ground; (c) the transmitter and receiver antenna gains in the ray launch and arrival directions. An area coverage diagram is made simply by making multiple point-to-point calculations with the transmitter location fixed and the receiver locations ranging over a two-dimensional grid. A (coherent field) coverage diagram corresponding to Figure 4 is shown in Figure 5 for a frequency of 38 GHz.

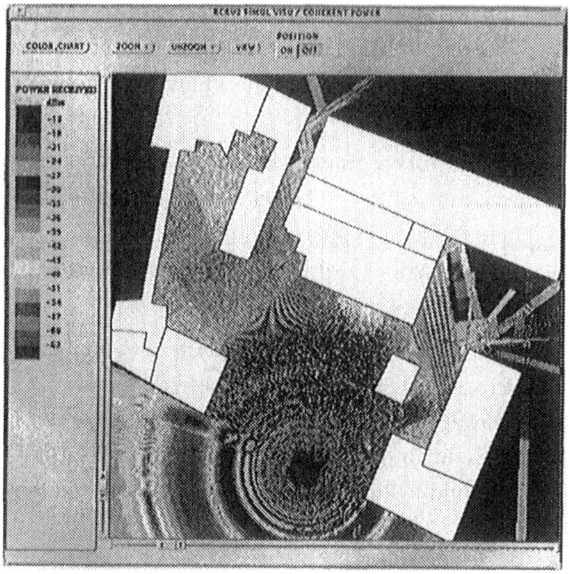

Figure 5: Coherent field coverage diagram.

Because the uncertainty of the positions of the buildings in a building database are likely to be large compared to a wavelength, and because of the limited level of detail in the database, the ***absolute*** phase of each ray will not generally be accurately known (although the ***relative*** phase of rays at two neighbouring locations will be). Thus although a coherent signal coverage plot (such as Figure 5) gives a good feel for signal variability, the value of received power at any given point is unlikely to be accurate. It is often more useful to show mean power in an area, together with an estimate of the signal variability in that area. The RMS power is often used, obtained by summing ray *powers* rather than ray ***amplitudes***.

Other quantities of interest are obtained directly from the ray parameters:

- power delay profile is obtained from the ray amplitudes and path lengths, and is closely related to the channel impulse response;
- the overall system impulse response is obtained from the channel impulse response (derived from the ray amplitudes and path lengths), convolved with the system impulse response;
- RMS delay spread is obtained by calculating the second central moment of the power delay profile;

– ray direction-of-arrival information can be very useful in identifying potential sources of interference, or in designing multipath countermeasures.

Site specific ray trace models can be two- or three-dimensional. For those applications where both transmitter and receiver antennas are low and the effects of propagation over the roofs of buildings can be ignored, a two-dimensional (plan view) ray trace may be adequate. If the application uses roof height antennas, or roof diffraction is important (for example in satellite-mobile studies), then a three-dimensional ray trace is required. Since the geometrical calculations are significantly more expensive in a full three-dimensional geometry, savings can be made in three-dimensional models by restricting building facets to vertical and horizontal surfaces (as in Figure 4).

In all ray tracing models, the time required for an area coverage calculation is simply proportional to the time for a point-to-point calculation. The grid resolution required clearly depends on the use to which the results will be put. A practical design tool will need far less resolution than shown in Figure 5, since coherent effects do not have to be resolved.

A numerical ray-based model must include all *significant* "rays" that represent each physical mechanism needed in the application, and contain models for the interaction of these rays with the buildings. The word "significant" is important as there is no upper limit in principle to the number of rays that can connect transmitter and receiver. What counts as "significant" will depend on the application of the model, but there will be some "dynamic range" below which the contribution of a ray will have no practical effect on system performance. If for example the dynamic range were 35 dB, and the "average" loss on reflection of a reflected ray were 10 dB, then all orders of reflected ray up to third would have to be accounted for, but fourth and higher order reflections could be ignored, assuming that a direct line-of-sight path exists between transmitter and receiver. Models differ in the methods they use to assess whether rays are "significant". Some set an *a priori* limit on the order of each mechanism included, while others track the amplitude of a ray as it propagates, and rejects it once it has dropped below the threshold.

The most numerically intensive stage of any ray trace model is finding the ray paths connecting the transmitter and receiver location. Two different techniques, "image" and "shooting" methods, are used. Both have analogies in computer graphics rendering.

3.1.1 Image methods

The image method [*Ikegami et al*, 1991; *Rossi and Levy*, 1992; *Anderson*, 1993; *Lawton* and *McGeehan*, 1994; *Charriere* and *Craig*, 1995] replaces each incident ray on an "object" (such as a building face or edge) with a virtual image whose characteristics model both the geometry and the electrical properties of the object. For example, if the database contains n building faces, there will be n first order reflection images of the transmitter (one for each face), n^2 second order reflection images of the n first order images, etc.

Since the major part of the computational workload lies in the geometrical calculations required to calculate the image positions and to check for building obstruction, the computation time of this algorithm T_1 is:

$$T_1 \propto n \sum_{i=1}^{i=d} n^{i-1} \tag{7}$$

where d is the maximum order of reflection. The run time increases exponentially with the order of reflection!

Such a computational complexity precludes the use of such a naïve image method for all but the smallest problems (for example, indoor propagation where only a few facets are needed). Fortunately, the computational complexity can be greatly reduced by noting that in a typical urban situation, the majority of facets are invisible from any given facet, *irrespective of the direction of arrival of an incident ray* (either because of their orientation, or because of obstructions). Thus the computational complexity can be greatly reduced by pre-calculating a facet connectivity matrix for the database (this only has to be done once for each database). An even greater reduction in workload derives from the realisation that the "field of view" of each image gets smaller and smaller as the order of the image increases, and the number of potential obstructers to be checked therefore decreases as the order of reflection increases; again this information can be pre-calculated (for each new transmitter position). In the "average" urban situation, the sparsity of the connectivity matrices ensures that the run time T_2 is:

$$T_1 \propto n \sum_{i=1}^{i=d} f(i,n) \tag{8}$$

where $f(i,n)$ is a decreasing function of i which tends to a positive constant value as i increases and becomes independent of n for large n. The computation is still intensive but now the run time tends to be proportional to the order of reflection d instead of exponential.

3.1.2 Shooting methods

In shooting methods [*Koushik et al*, 1995; *Luebbers* and *Schuster*, 1995], a large number of rays are launched from the transmitter in predetermined directions that uniformly cover a sphere (in three dimensions). Each ray is traced taking into account the mechanisms (specular- and non-specular reflection, diffraction, etc.) to be modelled, until the ray trajectory either "passes close to" the receiver, or leaves the area of interest. Since it is extremely unlikely that a ray launched in an arbitrary direction will intercept the receiver exactly, it is necessary to define a *reception sphere* around the receiver. Shooting methods rely on the continuity assumption that if a ray R_1 passes through the reception sphere, then there exists a ray path R_2 that connects the transmitter and receiver, such that the rays R_1 and R_2 have the same history (i.e. undergo the same interactions with building faces and edges, in the same order), and that all rays with launch angles between those of R_1 and R_2 also lie within the reception sphere.

This assumption is non-trivial and is not true in general since building edges will cause discontinuities in the ray trajectories. The more complex the building database, the greater the number of discontinuities that occur. If the reception sphere encompasses one of these shadow-boundary discontinuities, the assumption will fail. In order for the shooting method to work, the reception sphere must be made small enough to reduce the probability of mis-identifying a ray to an acceptable level.

The angular separation between the launched rays must also be chosen with care. If the rays are insufficiently dense, certain ray classes may be missed. A practical approach to reducing the probability of missed rays [*Luebbers* and *Schuster*, 1995] is to use a launch grid that is sufficiently dense to ensure that more than one ray of each ray class intercepts the reception sphere; alternatively the calculation can be re-run with an increased density of rays to check for convergence.

In shooting methods, the total received field is obtained by combining the individual ray parameters in the same way as for the image method, except that coherent results may need a careful consideration of phase to correct for the miss distance within the reception sphere.

An estimate of the computational time T_3 is:

$$T_3 \propto nd\,N \tag{9}$$

where N is the number of rays "shot". The factor nd accounts for the fact that each ray traced has to be checked for d interceptions with each of the n facets. As with the image method, the major part of the computational workload lies in the geometrical calculations required to check for building obstructions. Simple geometrical arguments suggest that $N \propto (R/r)^2$ where R is the maximum distance to which rays must be traced (i.e. the "size" of the database) and r is radius of the reception sphere. Again a facet connectivity matrix, as well as *a priori* limits on the ray launch angles, can be used to reduce the computational complexity. N is likely to be proportional to some power k of n (since each of R and r depend on n in this way), giving:

$$T_3 \propto n^{k+1} d \tag{10}$$

Comparing the image and shooting methods, the image method is guaranteed to find all rays of any class up to a predetermined order, while the shooting method has a finite probability of missing rays. However this can be made as small as desired at the expense of computation time. One of the strengths of the shooting method is that it allows the user to trade accuracy (in the sense of probability of missed rays) for speed. Both methods can be time-consuming for large databases, and it is not yet clear which is faster in each application area. It will be interesting to compare these two methods on the same problems and to carry out a more detailed complexity analysis. It seems likely that each method will have advantages, depending on the requirements of the problem. Indeed it may be that future models will include both methods of finding ray paths, switching between them depending on requirements.

3.2 Propagation mechanisms

As well as reflection and diffraction, other mechanisms may need to be included. Brief comments on implementation issues for ray trace models are now given.

Specular reflection: For flat surfaces that are large compared with the wavelength, Fresnel's laws are used to give the amplitude, phase, polarisation and direction of the reflected rays. The Fresnel reflection coefficients depend in principle on the electrical characteristics of the reflecting surface.

Since building surfaces are never truly smooth, it is important to make allowance for the effect of the surface roughness. This causes scatter into non-specular directions and a reduction of the energy in the specular direction. The classification of a surface as rough or smooth is wavelength- and incident angle-dependent, but a simple roughness reduction factor can be satisfactory for materials such as brick, concrete, grass and tarmac. At millimetric frequencies, where the surface roughness is comparable to the wavelength, the roughness reduction factor is more important than the electrical constants in determining the reflected ray amplitude.

It is questionable whether such simple roughness modelling is appropriate to account for the effects of larger scale features such as windows. Window and door frames (and other unmodelled architectural features such as stairwells and lift-shafts) can act as corner-reflectors, giving rise to a strong reflection in the direction of the incident ray.

Non-specular scatter: It is difficult to include the effects of the non-specular scattered signal within a ray trace model since each interaction with a building face is akin to a new source of "non-specular" rays in a shooting method ray trace, and the explosion in the number of rays makes the problem intractable. Some existing models [*Anderson*, 1993; *Koushik et al*, 1995] include a non-specular ray between the last building of a ray path and the receiver if no specular ray path exists. While giving a lower bound on the field strength at a receiver position that is not reachable by direct or specularly reflected rays, there is no theoretical reason why the *last* reflection should be the stage at which a non-specular scatter should be included rather than at any other stage.

It is worth mentioning that in computer graphics rendering, there is a method (radiosity) that is used to model diffuse sources and reflecting surfaces. It is not known whether radiosity methods have been applied to the radiowave propagation problem, and this may be a fruitful area of future study.

Diffraction: Electromagnetic waves can reach the geometrical optics shadow region behind a building by diffracting round, or over, the building. Even for locations where specular rays are present, diffraction can make an important contribution near geometrical optics shadow boundaries.

The conventional way of including diffraction is by means of the Geometrical or Uniform Theories of Diffraction (GTD/UTD) [see for example, *James*, 1986]. Extra "edge-diffracted" rays are introduced: each ray that is incident on an edge gives rise to a cone of diffracted rays. Calculation of the diffracted ray paths and the amplitudes of the diffracted rays is more complicated than for reflected rays, but the solutions are well-known and easily implemented.

GTD and UTD are corrections to geometrical optics, and so are good solutions *when the problem is quasi-optical.* An alternative approach at *low frequencies* (when the building sizes are comparable to a wavelength) is to describe the interaction of the electromagnetic wave with a building facet by means of physical optics (PO). In this approach, the "rays" simply interconnect the centres of the facets, the "reflection coefficients" of the "non-specular" reflections being determined by the "antenna pattern" of the facets. In this way the specular reflection and edge diffraction are accounted for at the same time. A computational advantage of this method compared to the GTD/UTD approach is that most of the ray paths are determined by the database independently of the transmitter or receiver positions. At high frequencies, the PO approach becomes numerically difficult. PO does not seem to have been widely adopted in site specific propagation modelling.

Building penetration: Electromagnetic waves can reach the geometrical optics shadow region behind a building by penetration through the building. This can be important at the higher (microwave and millimetre wave) frequencies, where diffraction edges are sharp. It is often difficult to predict the effects of building penetration because the materials, internal construction details, furnishings, etc. of the buildings are unlikely to be known. Even if they are, a full ray trace *within* the building itself would be numerically intractable. (Of course ray trace models are used for in-building propagation; our comment here relates to propagation where many hundreds of buildings may be present). At the higher frequencies, the attenuation by wall materials is high; in this case the penetration is concentrated at windows, and the field behind the building may be a strong function of position. Again the level of detail required in the building database to support small-scale predictions is unlikely to exist, and a simple empirical model of the "mean" attenuation of the building as a function of width can be used.

3.3 Other methods

Ray methods are not the only candidates for site specific modelling of urban propagation. The PE methods discussed in the previous section are beginning to be applied to the urban scatter/diffraction problem, both in two and three dimensions [*Levy*, 1993]. However these developments are still at an early stage. The modelling of high angle reflections and non-specular effects are not straightforward. Computation times are still high, although much effort has gone into improving this in recent years. Compared to the point-to-point approach of ray tracing, a PE calculation is intrinsically point-to-area (2D models) or point-to-volume (3D models). This is inevitably more expensive if a point-to-point calculation is required, although the comparison is not clear-cut if an area coverage is required (but it should be added that 2D PE point-to-area coverage diagrams are currently in the vertical plane, rather than in the horizontal plane).

A disadvantage of PE methods for application to wideband systems is that wideband effects are modelled by an explicit calculation of the channel transfer function in the frequency domain, multiplying computation times by a factor corresponding to the number of frequency components of interest. In a ray trace approach, the rays contain all the path delay, amplitude and phase information required to derive the channel transfer function directly. However a big advantage of PE methods is that they can deal in a uniform way with the effects of the atmosphere as well as scatter and diffraction from the terrain and buildings.

3.4 Building databases

The utility of any site specific propagation model clearly depends on the availability and quality of building databases. It has to be said that, with a few exceptions, such databases are at an early stage of development.

Two-dimensional (plan view) databases are relatively easy to obtain in countries with well-established mapping agencies. Large scale maps of town and cities are often available that contain street and building outlines, often already in digital form. However, maps can be of limited use unless they are up-to-date.

Three-dimensional building data are much less widely available, and so more costly to obtain. One method is to obtain a two-dimensional database as above, and to add the height information by surveying. For some applications, very accurate heights may not be needed, and a figure based on the number of floors may be adequate. In some countries, accurate, widespread stereographic air surveys of built-up areas have been carried out, and high quality 3D building databases can be obtained from these. At present this is a costly process involving skilled human interpreters, but given the demand there is hope that much of this work could be automated.

Architecture is one field where 3D building databases have been in use for some years. These can be very detailed. Figure 6 shows a part of such a database of the city of Glasgow, developed by the ABACUS group at the University of Strathclyde (M. Grant, personal communication, 1990), from which the subset shown in Figure 4 was taken. The level of detail shown in Figure 6 is more than is needed in current propagation models.

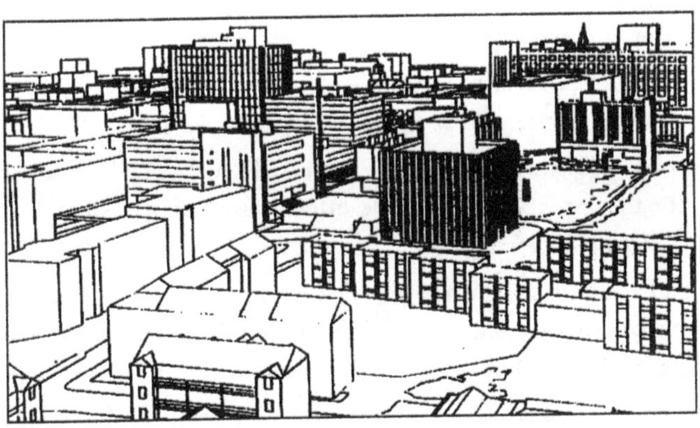

Figure 6: Example of a three-dimensional architectural building database

In order to encourage the wider development of building databases, it is desirable for propagation engineers to agree on the requirements (such as resolution, accuracy, level of detail, building materials) and data interchange standards (such as CAD). This would also make it easier for different groups to exchange data for model comparison purposes.

4 Conclusions

In this chapter, two areas of current research have been used to illustrate the way that numerical models are making an impact on propagation modelling. The advent of powerful desk-top computers and increasing availability of high quality environmental data have led to the evolution of theoretical techniques into:

- planning tools, being used in systems design;
- operational tools, being used for real-time forecasts;
- simulation tools, enabling parametric studies to be made;
- educational tools, providing an insight into the underlying physical processes.

The future for numerical methods in propagation is bright. The increasing power of hardware will enable more complex models and higher resolution data to be assimilated, and these propagation tools will become more widely available to the non-specialist user.

References

Anderson, H. [1993], "A ray-tracing propagation model for digital broadcast systems in urban areas", *IEEE Trans. on Broadcasting*, 39, pp 309–317.

Arora, R.K. and Wait, J.R. [1978], "Refraction theories of radio wave propagation through the troposphere–a review", *Radio Science*, 13, pp 599–600.

Barrios, A.E. [1994], "A terrain parabolic equation model for propagation in the troposphere", *IEEE Trans. Antennas and Propagation*, 42, pp 90–98.

Baumgartner, G.B., Hitney, H.V. and Pappert, R.A. [1983], "Duct propagation modelling for the integrated-refractive-effects prediction system (IREPS)", *IEE Proc. Part F*, 130, pp 630–642.

Booker, H.G. and Walkinshaw, W. [1946], The mode theory of tropospheric refraction and its relation to wave-guides and diffraction", in *Meteorological Factors in Radio-Wave Propagation*, Physical Society, London, pp 80–127.

Budden, K.G. [1961], *The wave guide mode theory of wave propagation*, Logos Press.

Charriere, P. and Craig, K.H. [1995], "A ray-based propagation tool for digital systems", *AGARD CP-574*, pp 6.1–8.

Cho S.H. and Wait, J.R. [1978], "Electromagnetic wave propagation in a laterally nonuniform troposphere", *Radio Science*, 13, p 253.

Craig, K.H. and Kennedy, G.R. [1987], "Case study of line-of-sight microwave multipath propagation", *Electron. Lett.*, 23, pp 483–484.

Craig, K.H. [1988], "Propagation modelling in the troposphere: parabolic equation method", *Electron. Lett.*, 24, pp 1136–1139.

Craig, K.H. and Levy, M.F. [1991a], "Parabolic equation modelling of the effects of multipath and ducting on radar systems", *IEE Proc. F*, 138, pp 153–162.

Craig, K.H. and Levy, M.F. [1991b], "A PC-based microwave propagation forecasting model", Seventh Int. Conf. on Antennas and Propagation, *IEE Conf. Pub.* 333, pp 523–526.

Dalrymple, R.A. and Martin, P.A. [1992], "Perfect boundary conditions for parabolic water-wave models", *Proc. R. Soc. London A*, 437, pp 41–54.

Dockery, G.D. [1988], "Modeling electromagnetic wave propagation in the troposphere using the parabolic equation", *IEEE Trans. Antennas and Propagation*, 36, pp 1464–1470.

Fournier, M. [1989], "Etude de la propagation dans une atmosphere inhomogene dans les directions horizontale et verticale par la methode de l'equation parabolique", *AGARD CP-453*, pp 21.1–13.

Hardin, R.H. and Tappert, F.D. [1973], "Application of the split-step Fourier method to the numerical solution of nonlinear and variable coefficient wave equations", *SIAM Rev.*, 15, p 423.

Hartree, D.R., Michel, J.G.L and Nicholson, P. [1946], "Practical methods for the solution of the equations of tropospheric refraction", in *Meteorological Factors in Radio-Wave Propagation*, Physical Society, London, pp 127–168.

Herring, R.N. [1985], "General raytracing techniques for predicting fields in an arbitrarily varying troposphere", Fourth Int. Conf. on Antennas and Propagation, *IEE Conf. Pub.* 248, pp 279–282.

Hitney, H.V. [1992], "Hybrid ray optics and parabolic equation methods for radar propagation modeling", Proc. RADAR 92, *IEE Conf. Pub.* 365, pp 58–61.

Ikegami, F., Takeuchi, T. and Yoshida, S. [1991], "Theoretical prediction of mean field strength for urban mobile radio", *IEEE Trans. Antennas and Propagation*, 39, pp 299–302.

ITU-R [1994], *Propagation in non-ionized media*, PN Series Volume, International Telecommunication Union, Switzerland, ISBN 92-61-05271-1.

James, G.L. [1986], *Geometrical theory of diffraction for electromagnetic waves*, Peter Peregrinus Ltd, ISBN 0 86341 062 6.

Ko, H.W., Sari, J.W., Thomas, M.E., Herchenroeder, P.J. and Martone, P.J. [1984], "Anomalous propagation and radar coverage through inhomogeneous atmospheres", *AGARD CP-346*, pp 25.1–14.

Koushik, P.M., Rappaport, T.S., Ahmed, M. and Zhang, N. [1995], "SISP–a software tool for propagation prediction", *AGARD CP-574*, pp 53.1–9.

Kuttler, J.R. and Dockery, G.D. [1991], "Theoretical description of the parabolic approximation/Fourier split-step method of representing electromagnetic propagation in the troposphere", *Radio Science*, 26, pp 381–393.

Lawton, M.C. and McGeehan, J.P. [1994], "The application of a deterministic ray launching algorithm for the prediction of radio channel characteristics in small-cell environments", *IEEE Trans. Vehic. Tech.*, 43, pp 955–969.

Lee, D. and McDaniel, S.T. [1987], "Ocean acoustic propagation by finite difference methods", *J. Comp. & Math. with Appl.*, 14, pp 305–423.

Leontovich, M.A. and Fock, V.A. [1946], "Solution of the problem of propagation of electromagnetic waves along the earth's surface by the method of parabolic equation", *J. Phys. of the USSR*, 10, pp 13–24.

Levy, M.F. [1990], "Parabolic equation modelling of propagation over irregular terrain", *Electron. Lett.*, 26, pp 1153–1155.

Levy, M.F. [1993], "Diffraction studies for microcellular applications", Eighth Int. Conf. on Antennas and Propagation, *IEE Conf. Pub. 370*, pp 76–79.

Levy, M.F. [1994], "Fast PE models for mixed environments", *AGARD CP-567*, pp 8.1–6.

Levy, M.F. [1995] "Horizontal parabolic equation solution of radiowave propagation problems on large domains", *IEEE Trans. Antennas and Propagation*, 43, pp 137–144.

Luebbers, R and Schuster, J. [1995], "Site-specific propagation prediction methods for urban environments", *AGARD CP-574*, pp 41.1–12.

McArthur, R.J. [1992], "Propagation modelling over irregular terrain using the split-step parabolic equation method", Proc. RADAR 92, *IEE Conf. Pub. 365*, pp 54–57.

Marcus, S.H. [1991], "A generalised impedance method for application of the parabolic approximation to underwater acoustics", *J. Acoust. Soc. Am.*, 90, pp 391–398.

Marcus, S.H. [1992], "A hybrid (finite difference–surface Green's function) method for computing transmission losses in an inhomogeneous atmosphere over irregular terrain", *IEEE Trans. Antennas and Propagation*, 40, pp 1451–1458.

Rossi, J.-P. and Levy, A.J. [1992], "A ray model for decimetric radiowave propagation in an urban area", *Radio Science*, 27, pp 971–979.

Shkarofsky, I.P. and Nickerson, S.B. [1982], "Computer modeling of multipath propagation: review of ray-tracing techniques", *Radio Science*, 17, pp 1133–1158.

Wait, J.R. and Spies, K.P. [1969], "Internal guiding of microwaves by an elevated tropospheric layer", *Radio Science*, 4, pp 319–326.

Webster, A.R. [1982], "Ray path parameters in tropospheric multipath propagation", *IEEE Trans. Antennas and Propagation*, 30, pp 796–800.

THE EQUATORIAL IONOSPHERE AND RADIO COMMUNICATIONS

B. Madhava Reddy

1 Introduction

The primary objective of the international union of Radio Science (URSI) has always been stimulating research on radio and telecommunications. The physics and propagation aspects of ionosphere were kept integrated until the fifties largely by stalwarts like Appleton. But with the considerable expansion in ionospheric research, especially after the International Geophysical year (1957-58), it tended to split distinctly as the propagation and physical aspects diverged [*Rishbeth*, 1994]. In fact we now have at least three specific streams: (i) structure and morphology to aid modelling (ii) propagation research using models as inputs (iii) physics which includes plasma aspects, dynamics and coupling mechanisms that generate inputs to the modellers. In addition, of course is the vast area of solar terrestrial predictions which encompasses all the above streams. In recent years, the ionospheric propagation research went through an intensive deemphasis, partly due to an assumption that we know what all there is to know and partly due to availability of better and more reliable systems such as satellite communication transponders. However the fact is the HF spectrum in most of the regions is getting more crowded, indicating that its utilisation is perhaps on the increase, at least in low latitudes. As conclusively validated in the most recent large scale war, "Desert storm operation", the cutting edge in communications was provided by ionosphere-reflected HF systems. To quote the chief of communications for Desert storm "HF had an important war-fighting and supporting role in the operations. In certain areas, such as reconnaissance and close air support, it was absolutely critical to mission success. In particular, NVIS (Near Vertical Incidence Systems) supports current maneuver warfare doctrine and ties the forces together". Thus, while there is little doubt about the continued use of ionospheric communications, the question that is often asked is if one can improve it any further by investing in more intensive research. The focus of this tutorial will be on the tropical ionosphere, with some emphasis on radio commumications and a peep into the future needs/trends in this area.

To quickly recapitulate some of the major shifts in ionospheric research trends, the space based research took off with a bang in the early sixties with the launch of satellites, particularly the topside sounder satellites. Subsequent developments in radars and signal processing brought back some glory to the ground based research through incoherent scatter technology. Incoherent scatter techniques allowed study of many more ionospheric parameters compared to conventional ionosondes, whether ground-based or space-based. The neutral atmosphere that hosts the ionosphere (thermosphere) is no more impervious thanks to some recent techniques such as Fabry-perot interferometer that has been used for the study of thermospheric temperature and winds. The sophisticated digital ionosondes that have emerged can provide the kind of detailed time structure of the ionosphere that may be required for some of the future systems.

2 The Equatorial Ionosphere - Thermosphere system:

2.1 The wind pattern in the tropics

An appreciation of the structure and dynamics of the equatorial and low latitude ionosphere thermosphere system is essential to properly integrate the objectives of this tutorial. A good number of published reviews related to this area make the job much easier [for example, *Killeen*, 1987; *Rishbeth*, 1977; *Titheridge*, 1995; *Abdu*, 1992] and this article has liberally drawn upon some of their material. The solar heating, mainly below 100 km drives zonal winds along with the E-region plasma across the horizontal magnetic field lines (the equatorial electrojet, EEJ) and the consequent EXB lifting of ionisation and the so called fountain effect produces the well known equatorial ionisation anomaly (EIA). This has been a subject of rather extensive observational and theoretical modelling studies [*Appleton*, 1946; *Rastogi*, 1959; *Abdu et al.*, 1990 and many others] and will not be discussed here except to mention about the large horizontal gradients produced in the tropical F-region, especially during day time. The consequences of such gradients will be a subject of further discussion later. It is presently accepted that a reasonable modelling has been achieved now for an undisturbed ionosphere with the available neutral atmospheric models, EUV data, ionisation cross sections and diffusion parameters [*Titheridge*, 1995 and the references there in]. However, the same is not true when it comes to our ability to develop consistent electric field and wind parameters, especially during certain periods, both quiet and disturbed. Since excellent reviews are already available on ionospheric and thermospheric winds, the discussion here will be limited to only those low latitude features that cause large changes in the ionospheric personality, posing serious problems to communication predictions. *Hedin and Mayr* [1973] reported a very interesting observation from the neutral mass spectrometer measurements on OGO-6 satellite during afternoon hours of a high solar epoch period. They reported that the latitudinal distribution of N2 and O densities show distinct maxima around ±20 degrees dip latitude with minimum at the dip equator, resembling the Equatorial Ionisation anomaly. With the aid of a quasi 3-dimensional model, they showed that the high concentration of ions at the anomaly peak impedes, through ion drag, the free flow of zonal winds that normally transport the heat away from the hot afternoon sector. Subsequently, *Raghava Rao et al.*, [1991] showed how the EIA modulates the zonal wind velocities and thermospheric temperature from the Dynamic Explorer (DE-2) wind and temperature spectrometer data. In a later paper [*Raghava Rao et al.*, 1993] the authors reported evidence for the presence of vertical winds in the thermosphere. Fig. 1 gives a summary of the zonal wind, neutral temperature and vertical wind features (from DE-2 data) coincident with EIA features during a single pass on April 17, 1982 at 2100 LST around 27 W longitude. While the argument that the neutral temperature peaks are co-located with the ionisation peaks, because of the larger ion-neutral drag impeding winds that transport energy, seems reasonable and perhaps even convincing, it may not be the only factor. It is well known now from satellite measurements that the F-region plasma temperatures show a decrease whenever there is a steep spatial increase in plasma concentration perhaps due to increased collisional cooling. Thus, the energy lost by the plasma may be credited to the neutrals. The physics of thermospheric heating from photo electrons is rather complex and the heating efficiency is only about 5 percent, with most of the energy going into space through air glow emissions from atomic and molecular states excited through inelastic collisional processes [*Killeen*, 1987]. In the lower thermosphere a good part of the potential energy of the ions is lost to lower altitudes during the dissociation of 02. That may not be true above 300 km where the ions will deliver most of the energy acquired from the electrons to the local thermosphere. The vertical winds depicted in the bottom panel are also consistent with the temperature peaks at the EIA crests. From other considerations not relevant here, it was estimated that the real zero in the vertical winds scale is really at- 11 m/s; thus, the derived upward velocity at ionisation crests is 15 m/s while the downward wind at the trough is 11 m/s.

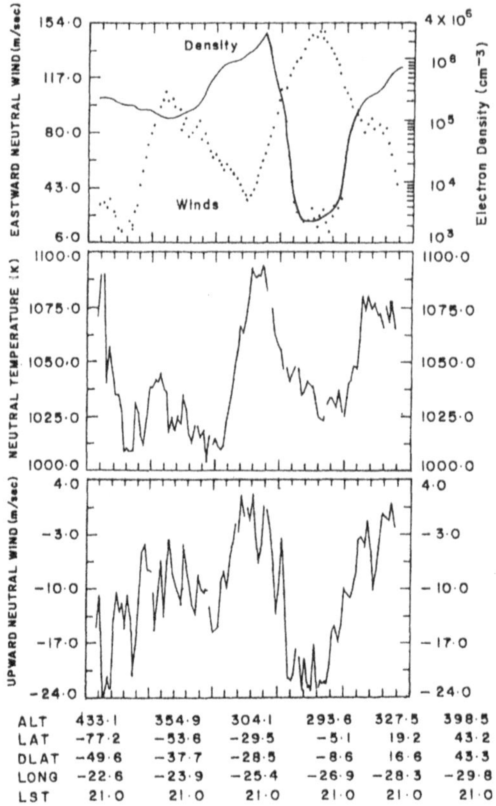

Fig 1. Dynamic-Explorer-2 data of electron density and zonal winds (upper panel), neutral temperature (middle panel) and vertical winds (lower panel) on April 17, 1982 at 2100 hrs LST around 27° W longitude [after Raghava Rao et al., 1993].

2.2 Thermospheric winds due to high latitude heat inputs

The distribution of the F-region electron densities are greatly manipulated by meridional thermospheric / ionospheric winds. It is possible that these winds may be the primary culprit in causing the frustratingly large day-to-day variability in the cardinal F-region parameters (critical frequency and height factors) which are not obviously related to any specific solar or magnetic event. In view of their importance for communication predictions, we will take a closer look at these meridional winds.

Winds in the Ionosphere -thermosphere system (ITS) are driven in response to pressure gradients set up by temperature differences. Above 150 Km, the major heat source in the tropics is the solar EUV maximising around 1400 hours local time. The second major heat source comes from high latitudes. Firstly, we have the large scale convection of the magnetic field lines due to solar wind interaction that can result in building up potential differences across the polar cap of the order of 35 kV to 120 kV depending on the level of disturbance [*Ahn et al.*, 1992]. It produces the Joule or frictional heating due to differential ion and neutral motions as the electric fields drive the plasma. In fact, the ion temperatures here far exceed the electron temperatures due to ions being forced past the neutrals [*Lakshmi and Reddy*, 1988]. The best estimates available for this energy suggest a range of 0.3 to 1.5 x 10^{11} Watts [*Rees et al.*, 1985 and *Killeen* and *Roble*, 1986]. Then, we also have the heat source from precipitating particles, which comes from the collisional retardation of magnetospheric energetic particles as they enter the lower atmosphere. It is not yet clear if most of these particles are from the solar wind, or terrestrial particles accelerated by the solar wind energy, though the latter idea is gaining ground. Anyway, the estimates for the total global thermosphere heating due to precipitating particles is lesser than that due to Joule heating, being in the range of 0.05 to 0.6 x 10^{11} Watts [*Spiro et al.*, 1982]. It should also be noted that the particle precipitation heating happens at substantially lower altitudes and hence Joule heating will play a major role in F-region wind generation. The other significant source of heating comes from dissipation of planetary waves, tidal waves and gravity waves originating in the lower atmosphere and propagating upwards. There are several fundamental points worth noting before we look at some F-region effects imposed by the winds. Solar EUV spectrum is the main source for the day time heating of the ionosphere; the softer EUV originating from the chromosphere changes by only a factor of 2 during a solar cycle while the shortest EUV from corona changes by as much as 100 times or more. The solar wind induced heating which is omnipresent becomes the major nocturnal heating. While the total energy in the solar wind is lesser compared to that in EUV, it is significant for a very good reason. The cross section offered by the terrestrial system to intercept EUV is a function of the diameter of solid earth; but the diameter of the magnetosphere being many times larger than that of earth, the cross section to intercept solar wind energy is a hundred times larger. Also, during night time, the ion drag is much lesser which is conducive to higher values of equatorward nocturnal winds. Once the winds arrive at low latitudes, the horizontal nature of the magnetic field allows the free flow of neutral air along with the plasma in a horizontal direction. This is unlike the situation at mid latitudes where the horizontal neutral wind is hindered by the propensity of the plasma (with its ionic component coupled tightly to the neutrals) to move along the non-horizontal magnetic field lines. Thus, at low latitudes, especially during night, the winds are limited only by inertial and viscosity effects [*Harris and Mayr*, 1975 and *Blum and Harris*, 1975]. Some of the recent models such as HWM 90 [*Hedin et al.*, 1991] and the NCAR-TGCM model [*Forbes and Roble*, 1990; *Fesen et al.*, 1993] do include a variety of recent data sets and give better fits, but a lot of uncertainties remain. Quantification of heating due to high latitude magnetospheric convection is extremely complex; for example, in summer, there is more heat input to the summer pole and the increased ion densities yield enhanced drag resulting in stronger winds and much greater heating. A recent study by *Brekke and Kamide* [1996] suggests that in the polar region, the fluctuating electric fields may enhance the frictional heating rate far above the traditional Joule heating rate. This is particularly so for electric fields with fluctuating frequencies less than twice the ion gyrofrequency. During particular types of disturbances, these electric fields may dominate and increase the polar heat inputs to unusual levels bringing strong meridional winds even to the low latitude ionosphere.

3 Sample observations of nocturnal winds at a low latitude station

The two most cardinal parameters relevant to long distance HF communications are the peak electron density and height of the F-region. Low latitude F-region dynamics are generally controlled by Electric fields and neutral winds. Of particular interest is the nocturnal dynamics of the F-region, especially in the post sunset period. The rate of change of the minimum virtual height of F layer, h'F, obtained from ionograms can be used to estimate the vertical drift velocities [*Bittencourt and Abdu*, 1981; *Batista el*, 1986]. Even meridional winds have been estimated using ionosonde data [*Krishna Murthy et al.*, 1990]. At a low latitude station, electric fields, neutral winds and to some extent diffusion play significant roles in the F-region vertical drifts:

$$Vz = ExB/B^2 - U \cos I \sin I - W_D \sin^2 I \quad (1)$$

Where U is the neutral wind velocity, I is the dip angle and W_D is the diffusion coefficient. For a station like Waltair (17.7° N, 83.3° E, dip angle 20°), the effect of diffusion is too small. Fig. 2 shows the night time variation of vertical drift velocity estimated from observed h'F variations at Waltair [P.V.S. *Rama Rao* and *P.T. Jayachandran*, personal communication]. The features to note here are the large upward drift in the post-sunset period followed by a sudden reversal, followed by yet another reversal after 10 P.M. These curves being averages, the gradients are modest; the actual daily data show rather spectacular gradients in these reversals. The nocturnal F-region also displays very interesting picture of its peak electron densities.

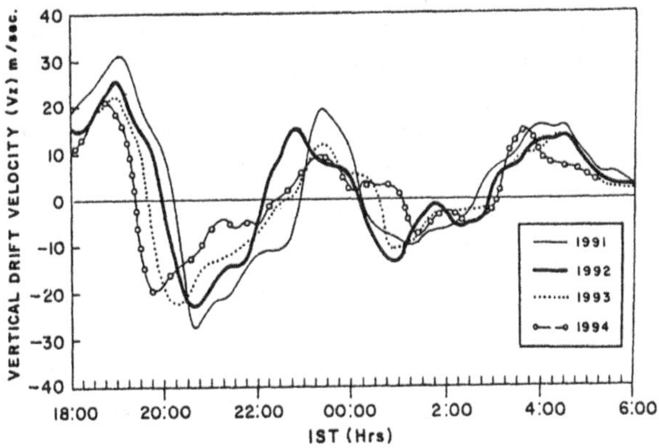

Fig. 2. Nocturnal variation of average vertical drift velocity during the equinoctial months of 1991, 1992, 1993 and 1994 at Waltair [Rama Rao and Jaya Chandran, personal communication].

Fig.3 shows a typical diumal variation of f_0F2 and h'F at waltair for modest solar activity equinox periods. It is very common to have f_0F2 values during most of the nocturnal hours higher than the daytime values,emphasizing the strong influence of vertical and meridional winds even for stations close to the dip equator [Also see *Anderson* and *Roble*, 1981]. Around 3 AM local time, h'F sharply increases (a common feature at waltair); but the f_0F2 decreases due to the limited plasma content in low latitude field tubes. This rapid collapse of F2 in the early morning hours abetted by increase in the height of the layer causes serious problems in HF communications, as both these factors decrease the Maximum useable Frequency (MUF).

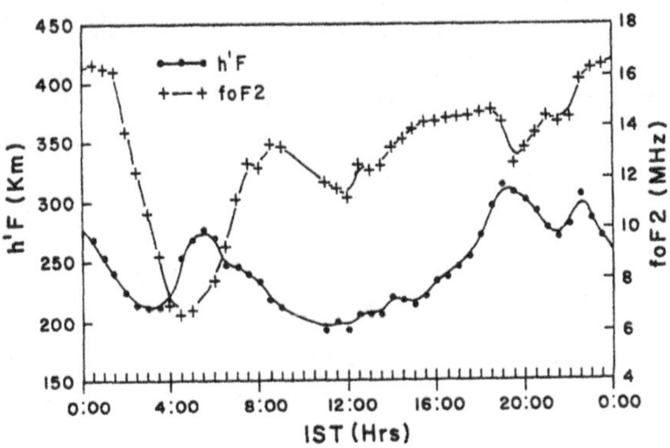

Fig. 3. Diurnal variation of h'f and f_0F2 on March 29, 1992 at Waltair showing high foF2 values during night and a steep fall in pre-dawn hours [Rama Rao and Jaya chandran, personal communication].

4 Case studies of unusual Equatorial Plasma depletions during quiet and disturbed periods

The data sets to demonstrate intense plasma depletions during northern summer at night were generated by the Indian SROSS-C satellite launched on May 20, 1992 by the Indian space research organisation from Sri Harikota near Madras. The lower-than-planned orbital parameters restricted the life of the satellite to less than 2 months, but provided a unique opportunity to study the low latitude ionosphere around an altitude of 250 km which is much below the normal satellite domain. The data was recorded in real time from the satellite over Bangalore (12.5° N, 77.3° E), Lucknow (26.8° N, 80.8° E) and over Mauritius (20° S, 56° E). This gave a wide latitudinal coverage in the tropics and the data reported here pertains to altitudes around 250 km with the altitude variation during any particular pass being with in ± 10 km. Fig. 4 shows the data sets which are essentially the latitudinal distribution of night time (2100 hrs to 2330 hrs LT) ion density (electron density) from the Retarding Potential Analyser experiment below the F-region peak on four different nights [*Garg et al.*, 1996]. Curve (1) in Fig.4 depicts the usual ion density distribution with latitude showing the EIA with a shift towards the summer hemisphere.

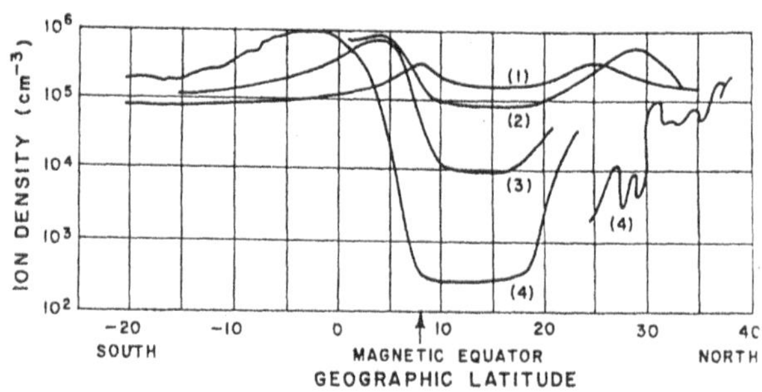

Fig. 4. The Indian SROSS-C satellite data on four nights during June-July 1992 showing latitudinal distribution of night time ion density in the bottom side F-region. Unusual plasma depletions north of the magnetic equator are associated with enhancements towards south [after Garg et al., 1996].

The ion densities in the southern hemisphere (10^5/cc) are about an order of magnitude less than their day time values while they are a few times higher in the northern (summer) hemisphere. On certain nights the ion densities collapse spectacularly from almost half-a-million per cc to as low as a mere 500 per cc (curve 4) north of the dip equator. It is also interesting to note that the larger the depletion north of the dip equator, the greater is the enhancement in ion density in the southern hemisphere. While a comprehensive discussion on these results is beyond the scope of this presentation, the following tentative arguments can be made. As it is summer in the north, strong trans-equatorial meridional winds can be expected - which can be extremely strong on certain nights. These neutral winds can literally sweep the plasma from north to south. In the south, close to dip equator, the upward winds will not let the plasma sink to lower altitudes which is conducive to large accumulation of plasma around 5-10 degrees south of the dip equator. Perhaps the whole phenomenon is magnified to some extent because of increase in h'F north of the equator on those nights-which means that the satellite is really in the valley below the F-layer. Ionosonde observations north of dip equator do support such a situation; but no ionosonde data is available south of the equator.

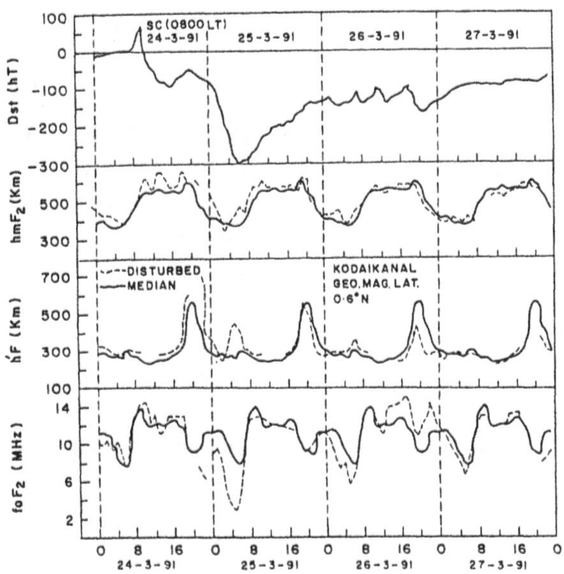

Fig. 5. Ionospheric data for a storm with a Dst excursion of 300 nT (top panel) in March 1991. Peak height (hm F2) of the F2-layer, base of the F-region height (h'f) and fo F2 are shown for the disturbed period (broken curve) along with monthy median values (continuous curve). The data pertains to Kodaikamal.

Very large plasma depletions similar to those discussed above, but at the peak of the F-region, have been observed in the nocturnal equatorial ionophere during severe magnetic disturbances. A large number of such severe storms with reference to their effects on the equatorial ionosphere were studied recently [*Lakshmi et al.*, 1996, also see *Lakshmi et al.*, 1991]. A single representative case will be described here as an example. Fig. 5 shows the details for a severe storm that occurred at 0300UT on 24 March 1991 with a well defined SC and with the maximum negative excursion of Dst reaching upto 300nT (top panel of Fig. 5). The other parameters shown are hmF2 (the peak height of the F2 layer), h'F (the base height of the F-layer) and f_0F2. These data pertain to *Kodaikanal* (Geographical Lat. 10° 14' and dip lat 0.6° N). The most striking feature to be observed is the dramatic collapse of F-region ionisation as evidenced by the foF2 values (broken curve) on the night of 24-25 March 1991 during the main phase of the storm. The sharp fall in f_0F2 values during 0100-0500 LT was rapid, with f_0F2 values collapsing from 9.5 MHz to 3.0 MHz while the peak heights (hmF2) in general were higher than the monthly median values by as much as 50-100 km. *Lakshmi et al.*, [1991] did consider the possible effects of enhanced neutral temperatures to explain such ionisation collapses, but the results were not convincing. It is in this context that *Lakshmi et al.*, [1996] based their arguments on meridional winds [*Krishna Murthy et al.*, 1990; *Krishna Murthy* and *Hari*, 1992] sweeping the plasma from north to south (Kodaikanal geographically is well in the northern hemisphere and equatorward winds can be expected even in Equinox). In fact, storm simulations using the UCL coupled ionosphere-thermosphere model showed cross equatorial winds driven by Joule heating under equinox conditions [*Fuller-Rowell et al.*, 1994]. Such meridional winds can drive the plasma along the near-horizontal magnetic field and physically transport it to points south of Kodaikanal. When such forced migration takes place, the depleted plasma can not be replenished either from below or from the top because the magnetic field lines are horizontal and there is no production, it being night time. Thus, as the plasma is swept over from Kodaikanal towards south, the base of the F-region (h'F) is also subjected to this sweep, in fact more effectively, because the ion-neutral coupling is all the more stronger in the lower F-region than at the peak. This should be the cause for the increase in h'F as observed from the ionogram.

5 Radio communications in the tropics

5.1 Ionosphere and the radio spectrum

Quite contrary to common belief, ionosphere exercises its influences on essentially all bands of Radio spectrum from the Extreme low frequencies (ELF) to as high as 10 GHz or even higher. ELF, VLF and LF used in Navigation, Maritime Communications, Time and Frequency Transmissions etc. propagate mainly as ground waves, but also through ionospheric reflection under certain conditions. MF used for short and medium range broadcasting gets extended range of coverage in the night time through ionospheric propagation. The HF spectrum which has been the prima donna until the sixties for long distance broadcasts and point-to-point links depends entirely on ionospheric support. The VHF spectrum is used mostly in line of sight (LOS) modes which include TV broadcasts and certain satellite services-except some scatter systems using either the troposheric irregutarities [*Reddy*, 1987], meteor trails or irregularities in the D-region of the ionosphere. Otherwise, ionosphere is a major nuisance to VHF systems. One is through the F-region irregutarities which degrade performance of satellite-based systems through the so called scintillations. The other is in TV broadcasts where the ionosphere, especially the sporadic E-layer, can support TV signal propagation to unintended target zones in far off areas, causing interference and serious problems in spectrum management. Satellite-based radio systems in UHF (300 MHz to 3000 MHz) aud even in SHF (3GHz to 30GHz) are not immune to ionospheric degradation.

5.2 HF communications in low latitudes

The use of HF spectrum with ionospheric support for broadcasting and point-to-point links is blessed with a number of positive factors in the tropics. Because of low solar zenith angles, we have simply greater solar ionising radiations per unit area leading to greater electron production. So we have higher electron densities facilitating higher operating frequencies with lesser ionospheric absorption.

Higher frequencies also yield better signal-to-noise ratio because of lower Atmospheric Radio Noise. Antenna dimensions and real estate requirements will also be modest at higher frequencies. The low latitude ionosphere is well protected by the terrestrial magnetic field from the invasion of solar particles which cause a host of particle effects including the Polar Cap Absorption events at high latitudes. Even the delayed effects of these particle events, like the ionospheric storms, do not cause major problems in low latitudes, except when they are very severe disturbances as discussed in section 4. HF services in the high and mid latitude zones are seriously dislocated during the main phase of the storm as the MUF values plunge far below the predicted medians. Some of the recent broad-casting systems can use real time data provided by state-of-the-art ionosonde networks [*Wilkinson et al.*, 1992] in frequency-agility links; but the main problem will remain spectrum management. However, there are also some specific problems peculiar to low latitudes. The spatial gradients associated with the Equatorial Anomaly can pose a variety of problems including asymmetric ray paths, deviation from great circle propagation etc. [*Oyinloye*, 1987; *CCIR*, 1986]. In addition, a peculiar situation of one way communication can arise as it happened for Delhi-Madras links, caused by these large horizontal gradients in electron densities [*Lakshmi et al.*, 1979]. Fig.6 shows how the MUF values depend upon the horizontal electron density gradients; the situation is worse for long distance circuits which will have to operate with large angles of incidence. For example, if the point of reflection of a north-south HF link is equatorward of the anomaly peak in the northern hemisphere, the radio waves incident on the ionosphere for the northward circuit will continuously encounter along the path increasing levels of electron density both due to vertical as well as horizontal gradients; however, for the southward path incident at the same point, the sense of the horizontal gradient is reversed. As a consequence, the real MUF values in the two opposite directions can vary by a large margin causing communication failure in one direction.

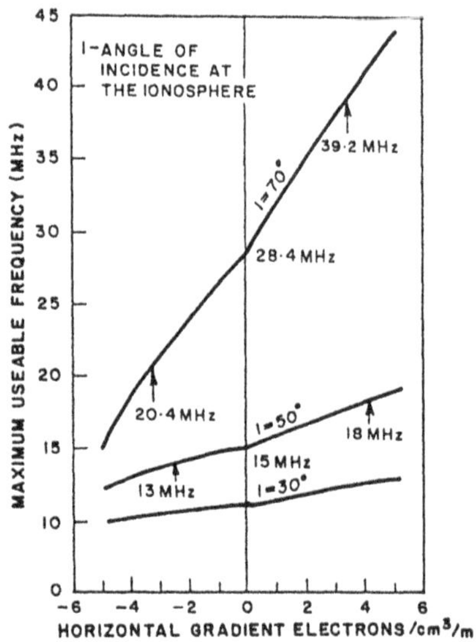

Fig. 6. Effects of horizontal electron density gradients on the maximum useable frequencies in HF communication paths for different angles of incidence.

The low latitude ionosphere also exhibits steep temporal gradients, particularly during the dawn and dusk transition periods. Fig.7 shows the pre-dawn collapse and the post dawn appearance of the F-region at Jicamarca and Kodaikanal [*Lakshmi et al.*, 1980]. It may be noted that the transitions are much more rapid and steep at Kodaikanal than at Jicamarca. Such steep pre-dawn changes at Kodaikanal have been attributed to a significant increase in the eastward ionospheric drift in the predawn hours before the morning reversal [*Chandra and Rastogi*, 1970; *Rastogi*, 1988]. This rather large eastward drift (or west ward electric field) would push down the F-layer, causing its collapse. Situation at places away from the dip equator is quite different, as discussed in section 2.2., especially if they are abetted by thermospheric winds. The consequence of such collapse of the layer is almost a failure of long distance HF links; this problem is particularly serious in the low solar epoch.

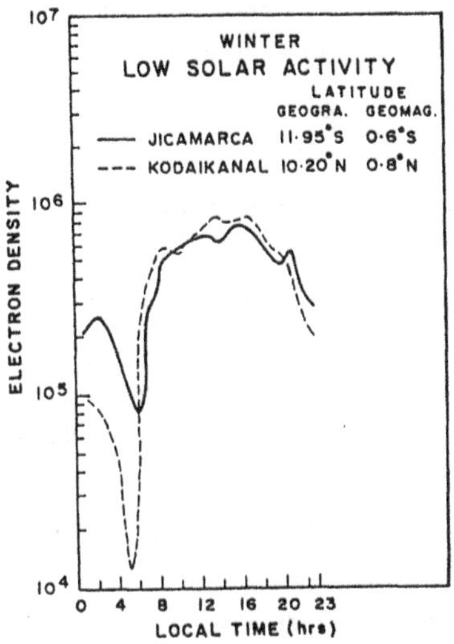

Fig. 7. Diurnal plots of F-region peak electron densities for two stations near the magnetic equator, Kodaikamal and Jicamarca.

The day-to-day variability in F-region parameters, apparently unrelated to any specific solar or magnetic event is perhaps the biggest challenge to ionospheric forecasters and the problem is particularly severe in low latitudes [*Lakshmi*, 1994]. There have been some attempts to generate statistical pictures of ionospheric variability using a large volume of data [*Aggarwal et al.*, 1979], but the problem persists. One consequence is that HF operators looking for higher reliability are forced to operate at frequencies much lower than the optimum working frequencies, thus crowding the lower part of the HF spectrum and also using higher powers due to higher ionospheric absorption and higher Atmospheric Radio Noise at lower frequencies. This variability, also known as ionospheric weather, is obviously a consequence of variable electric fields, thermospheric winds etc. and will be impossible to predict; dynamic adaptive systems may be the only solution with proper caution on spectrum management. In summary, HF has several advantages at low latitudes, in addition to being very simple and inexpensive. It is also particularly suitable to vast countries dominated by rural and scattered communities which abound in the tropical region.

5.3 Restrictions imposed by Ionosphere on Trans-Ionospheric propagation

The presence of the ionosphere in the propagation path between a satellite and the ground receiver gives rise to various effects that degrade the performance of the satellite based radio systems [*Klobuchar*, 1985]. The group path delay which represents the additional delay caused by the retardation due to the electrons in the ionosphere is given by

$$t = 40.3 / cf^2 \times TEC \quad \text{(in seconds)} \qquad (2)$$

where c is the velocity of light in free space in m/sec., f is the operating frequency in Hertz and TEC is the total electron content in a unit cross section column of one square metre area along the path. This will introduce significant errors in ranging systems. This is being reduced in systems with large bandwidths by using a multifrequency correction technique. The other problems include the Faraday polarisation rotation which is especially complex for mobile receiving systems; angular refraction which gives an error in elevation angles in tracking systems and distortion of pulse waveforms due to the dispersive nature of the ionospheric medium. But perhaps the most deleterious effect imposed by the equatorial ionosphere on system performance in trans-ionospheric propagation is the fluctuations in signal power and phase as a result of inhomogeneities in the refractivity caused by a moving ionosphere embedded with irregularities [*Aarons*, 1977; *Dabas and Reddy*, 1990; *Banerjee et al.*, 1992]. A recent study over India using data from two satellites during the years 1984 to 1991 collected at Dehli has produced a detailed morphology of scintillations at 4 GHz [*Dabas et al.*, 1991]. Engineers designing Earth-space links need to determine the fade margins necessary to overcome the degradation caused by scintillations.

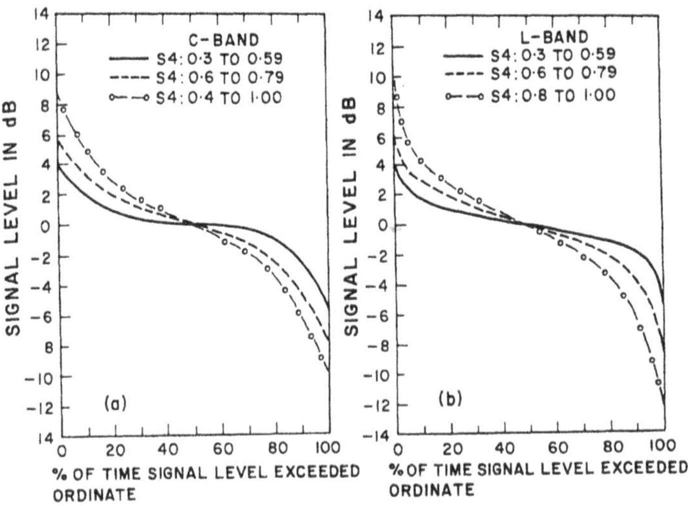

Fig. 8. Cumulative fade duration distribution for trans-ionospheric satellite signals in C and L bands for the three sets of S4 values. S4 is a measure of the intensity of scintillations.

Presently, an adhoc fade margin of 2 to 3 dB is allocated to compensate all propagation-related effects. Fig. 8 (a and b) from data at Delhi illustrates the inadequacy of such small margins for high reliability systems at C (4 GHz) and L (1.5 GHz) bands respectively. The index S4 is a measure of the intensity of scintillation and is defined as the square root of the variance of received power divided by the mean value of the received power. A detailed study of the recordings of telemetry transmissions at 4 GHz from the Indian geostationary satellite INSAT-1B (74° E) showed that during strong scintillations, the signal frequently faded beyond 10 dB relative to normal level. In digital transmission system, the worst acceptable probability levels are defined in terms of C/N (carrier-to-Noise ratio at the receiver input) degradation margin for a given data transmission rate. As an example, for a satellite-based computer data network using a 1.8 m terminal with coherent differential phase-shift keying (CDPSK) modulation system transmitting down link data at the rate of 153.6 K bits/s, the probability of error, P(e) at different scintillation fade depth levels has been computed up to a maximum of 12 dB (the worst level observed at Delhi). Table 1 shows the results of these calculations below a specified fade depth. These data pertain to the high solar activity year, 1989. It can be seen that the bit error rate probability is extremely sensitive to even small changes in fade levels; a change of 1 dB increases the P(e) by an order of magnitude.

Table 1 Probability of Bit Error Rate P(e) for different fade depths and their percentage of occurrence at 4 GHz in 1989

Fade Depth (dB)	Percentage of Nights the Fade depth exceeded levels in the first column	P(e)
0	-	1.14×10^{-6}
-2	33.33	1.40×10^{-4}
- 4	20.45	2.80×10^{-3}
- 6	12.31	1.90×10^{-2}
- 8	5.56	6.4×10^{-2}
- 10	1.25	1.3×10^{-1}
- 12	0.33	2.2×10^{-1}

6 Future scope and conclusion

While the ionosphere has been a subject of intensive study through radio and in situ techniques for several decades now, it is only recently that we have access to significant volume of data on thermosphere and its coupling to the plasma. Several types of high power radars, Fabry-Perot interferometers, improved satellite measurements, new development to derive thermospheric dynamics from conventional ionosondes and large computing facilities for global circulation modelling have given insight into ionosphere-thermosphere coupling and the processes through which energy dumped in the neutral atmosphere can manipulate the ionosphere, especially the tropical F-region. It may be possible in near future to use Fabry Perot technique for reliable day time measurements also [*Sridharan et al.*, 1994]. Traditional HF communications will continue to be important, especially in the tropics with new digital packages coming into vogue and this requires better understanding of the fine structure of the ionosphere through use of more sophisticated digital ionosondes. There is yet another HF system which may assume a critical role in the future - the HF over-the-Horizon Radar. As satellite and air-borne surveillance systems are vulnerable with the development of laser weapon systems, the HF radar which can be operated from deep inside the country is an attractive alternative. Several countries have these systems already operational [for example, see *Earl and Ward*, 1987]. Because of serious troposheric problems above 10 GHz, frequencies below 10 GHz will continue to play a dominant role in satellite based radio systems and all the ionosphere-imposed restraints on frequencies below 10 GHz will have to be studied and modelled to optimise these systems.

References

Aarons, J., [1977] "Equatorial Scintillations: A Review", *IEEE Trans. Antennas Propagation*, AP25, 729.

Abdu, M.A., [1992] "Equatorial ionosphere-thermosphere system (EITS): An overview of recent results", *Proceedings of STEP symposium*, John Hopkins University, August, 1992.

Abdu, M.A., G.O. Walker, B.M. Reddy, J.H.A. Sobral, B.G. Fejer, T. Kikuchi, N.B. Trivedi and E. P. Szuszezewicz [1990] "Electric field versus neutral wind control of the equatorial anomaly under quiet and disturbed conditions: a global perspective from SUNDIAL 86", Ann. *Geophysicae*, 8, 419-430.

Aggarwal, S., D.R. Lakshmi and B.M. Reddy [1979] "A simplified indexing of F-region geophysical noise at low latitudes" *Solar Terrestrial predictions proceedings* (Ed. R.F. Donnelly), Boulder, U.S.A.

Ahn, B.H., Y. Kamide, H.W. Krochl and D.J. Gorney [1992] "Cross polar cap potential difference auroral electro-jet indices and solar wind parameters", *J. Geophys. Res.*, *97*, 1345-1352.

Appleton, E.V [1946] "Two anomalies in the ionosphere", *Nature*, 157, 691.

Anderson, D.N. and R.G. Roble [1981] "Neutral wind effects on the equatorial F-region ionosphere", *J. Atm. Terr. Phys.*, 43, 835-844.

Batista, I.S., M.A. Abdu and J.A. Bittencourt [1986] "Equatorial F-region vertical plasma drifts: Seasonal and longitudinal asymmetries in the American Sector" *J.Geo Phys. Res.*, 91, 12055-12064.

Banerjee, P.K., R.S. Dabas and B.M. Reddy [1992] "C and L band transionospheric scintillation experiment: Some results for application to satellite radio systems", *Radio Science*, 27, 955-969.

Bilitza, D., [1991] "Electron and ion temperature data for ionospheric modelling", *Adv. Space*. Res. 11, 138-148.

Bittencourt, J.A. and M.A. Abdu [1981] "A theoretical comparison between apparent and real vertical ionisation drift velocities in the equatorial F-region" *J. Geo Phys., Res.*, 86, 2451-2454.

Blum, P.W. and I. Harris [1975] "Full non-linear treatment of the global thermosphere wind system II Results and comparison with observations", *J. Atm. Terr. Phys.*, 37, 213-236.

Brace, L.H. and R.F. Theis [1978] "An empirical model of the interrelationship of electron temperature and density in the day time thermosphere at solar minimum" *Geo Phys., Res.*, Lett., 5, 275-278.

Brekke, A and Y. Kamide [1996] "On the relationship between Joule and frictional heating in the polar ionosphere" *J. Atm. Terr. Phys.*, 58, 139-143.

CCIR [1986] Recommendations and Reports of CCIR, International Telecommunication Union, Geneva, Rep. 250-6.

Chandra, H. and R.G. Rastogi [1970] "Daily variation of F-region drifts at Thumba" *J. Atmos. Terr. Phys.*, 32, 1309-13 1 1.

Dabas, R. S. and B.M. Reddy [1990] "Equatorial plasma bubble rise velocities in the Indian sector determined from multistation scintillation observations", *Radio Science*, 25, 125-132.

Dabas, R.S., Banerjee, P.K., S. Bhattacharya, B.M. Reddy and J. Singh [1991] "Giga Hertz scintillation observations at 22.5 N magnetic latitude in the Indian Zone", *Radio Science*, 26, 750771.

Earl, G.F. and B.D.Ward [1987] "The frequency management system of the Jindalee over-the-horizon backscatter HF radar" *Radio Science*, 22, 275-291.

Fesen, C.G., R.G. Roble and E.C. Ridley [1993] "Thermospheric tides simulated by the National Centre for Atmospheric Research thermosphere ionosphere general circulation model at equinox" *J. Geophys.*, 98, 7805-7820.

Forbes, J.M. and R.G. Roble [1990] "Thermosphere-Ionosphere coupling: An experiment in interactive modelling", *J. Geophys. Res.*, 95, 201-208.

Fuller-Rowell, T.J., M.V. Codrescu, R.J. Moffet and S. Quegan [1994] "The response of the thermosphere and ionosphere to geomagnetic storms", *J. Geophys. Res.*, 99, 3893-3914.

Fuller-Rowell, T.J., D. Rees, S. Quegan, R.J. Moffet and G.J. Bailey [1987] "Interactions between neutral thermospheric composition and the polar ionosphere using a coupled ionosphere thermosphere model" *J. Geophys., Res.*, 92, 7744-7748.

Garg, S.C., P. Subrahmanyam, M. Bahl, H.K. Maini, P. Chopra, S.K. Singhal, T John, B.C.N. Rao and Y.V. Somayajulu [1996] "Incidence of extensive plasma density depletions in the lower F-region", *Adv. Space Res.*,18, 103-106.

Harris, I. and H.G. Mayr [1975] "Diurnal variations in the thermosphere", *J. Geophys. Res.*, 80, 3925-3933.

Hedin, A.E., M.A. Biondi, R.G. Burnside, G. Hernandez, R.M. Johnson, T.L. Killeen, C. Mazaudier, J.W. Meriwether, J.E. Salah, R.J. Sica, R.W. Smith, N.W. Spencer, V.B. Wickwar and T. S. Virdi [1991] "Revised global model of thermospheric winds using satellite and groundbased observations", *J. Geophys. Res.*, 96, 7658-7688.

Hedin, A.E. and H.G. Mayr [1973] "Magnetic control of the near equatorial neutral thermosphere" *J. Geophys. Res.*, 78, 1688-1691.

Killeen, T.L. [1987] "Energetics and dynamics of the earth's thermosphere", *Rev. Geophys.*, 25, 433-454.

Killeen, T.L. and R.G. Roble [1986] "An analysis of the high latitude atmospheric wind pattern calculated by a thermospheric general circulation model, 2, Neutral parcel trajectories" *J. Geophys. Res.*, 91, 11291-11307.

Klobuchar, J.A. [1985] Ionospheric Radio Wave propagation, section 10.8, Handbook of Geophysics and space environment (Ed. A.S. Jursa), *Air Force Geophysics Lab Publication.*

Krishna Murthy, B.V. and S.S. Hari [1992] "Equatorial night time thermospheric meridional winds: Seasonal variations and magnetic activity effects", *proceedings of STEP Symposium*, Japan.

Krishna Murthy, B.V., S.S. Hari and V.V. Somayajulu [1990] "Night time equatorial thermospheric meridional winds from ionospheric h'F data", *J. Geophys. Res.*, 95, 4307-4310.

Lakshmi, D.R., S. Aggarwal P.K. Pasricha and B.M. Reddy [1979] "HF communication problems at low latitudes due to steep spatial and temporal electron density gradients", *Indian J. Radio and Space Phys.*, 8, 101-105.

Lakshmi, D.R., S. Aggarwal and B.M. Reddy [1980] "On the large spatial and local time electron density gradients in the equatorial F-region", low latitude aeronomic processes, Advances in Space exploration, COSPAR Sym. S-8 (Ed. A.P. Mitra], Pergamon Press.

Lakshmi, D.R., B.C.N. Rao, A.R. Jain, M.K. Goel and B.M. Reddy [1991] "Response of equatorial and low latitude F-region to the great magnetic storm of 13 March, 1989" *Ann. Geophysicae*, 9, 286-290.

Lakshmi, D.R., and B.M. Reddy [1988] "On the anomalous ion temperatures observed at high latitude ionization troughs by AE-C satellite", *Indian J. Radio and Space Phys.*, 17, 89-92.

Lakshmi, D.R. [1994] "Tropical ionosphere and Radio communications", *Proc. Indian National Science Academy*, 60 A, 243-262.

Lakshmi, D.R., B. Veenadhari, R.S. Dabas and B.M. Reddy [1996] "Sudden post midnight decrease of equatorial F-region electron densities associated with severe magnetic storms", *communicated to Ann. Geophysicae.*

Oyinloye, J.O. [1987] "Equatorial lonosphere", Handbook on Radio propagation for Tropical and Sub-Tropical Countries, Edited by A.P. Mitra, B.M. Reddy, S.M. Radicella, J.O. Oyinloye and S. Feng, published by URSI.

Raghava Rao, R., W.R. Hoegy, N.W. Spencer and L.E. Wharton [1993] "Neutral temperature anomaly in the equatorial thermosphere - A source of vertical winds" Geophys., Res., Lett., 20, 1 023- 1 026.

Raghava Rao, R., L.E. Wharton, N.W. Spencer, H.G. Mayr and L.H.Brace [1991] "An equatorial temperature and wind anomaly (ETWA)", *Geophy. Res. Lett.*, 18, 1193-1196.

Rastogi, R.G. [1988] "Collapse of the equatorial ionosphere during the sunrise period" *Ann. Geophysicae*, 6, 205-210.

Rastogi, R.G. [1959] "The diurnal development of the anomalous equatorial belt in the F2 region of the ionosphere", *J. Geophys., Res.*, 64, 727-732.

Reddy, B.M. [1987] "The physics of the troposphere", Handbook on Radio propagation for tropical and sub-tropical countries, Edited by A.P. Mitra, B.M. Reddy, S.M. Radicella, J.O. Oyinloye and S. Feng, published by URSI.

Rees, D., R. Gordon, T.J. Fuller-Rowell, M. Smith, G.R.Carignan, T.L. Killeen, P.B. Hays and N.W. Spencer [1985] "The composition, structure, temperature and dynamics of the upper thermosphere in the polar regions during October to December, 1981", *Planet. Space Sci.*, 33, 6 1 7-666.

Rishbeth, H. [1977] "Dynamics of the equatorial F-region", *J.Atm. Terr. Phys.*, 39, 1159-1168.

Rishbeth, H., [1994] "What became of Appleton's ionosphere" *J. Atm. Terr. Phys.*, 56, 713-725.

Spiro, R.W., P.H. Reiff and L.J. Maher, Jr., [1982] "Precipitating electron energy flux and auroral zone conductance: An empirical model", *J. Geophys. Res.*, 87, 8215-8227.

Sridharan, R., D. Pallam Raju, R. Raghava Rao and P.V.S. Rama Rao [1994] "Precursor to equatorial spread-F in OI 630.0 nm dayglow" *Geophys. Res. Lett.*, 21, 2797-2800.

Titheridge, J.E. [1995] "Winds in the ionosphere-A review", *J. Atm. Terr. Phys.*, 57, 1681-1714.

Wilkinson, P.J., C. Beveins, R.A. Smith, C. Yuile, G. Robinson, Y. Gin and G. Patterson [1992] "The IPS 5A ionosonde network" Solar-Terrestrial Predictions-IV, *Proceediugs of STP workshop, Canada* (Ed.J. Hruska, M.A. Shea, D.F. Smart and G. Heckman), Vol.3, 585-594.

RADIO EMISSION FROM INSTABILITIES IN SPACE PLASMAS: MARGINAL STABILITY, STOCHASTIC GROWTH AND FINE STRUCTURES

Donald B. Melrose

1 Introduction

Plasma instabilities can lead to radiation from a space or an astrophysical plasma either indirectly or directly. Two specific such processes are discussed here: plasma emission, which is an indirect emission process first discussed by *Ginzburg* and *Zheleznyakov* [1958], and electron cyclotron maser emission (ECME), which is a direct emission process first discussed in the presently favored form by *Wu* and *Lee* [1979]. Plasma emission is the accepted emission mechanism for most solar radio bursts, notably type III bursts, for analogous emissions from planetary bow shocks and from the auroral ionosphere. ECME is the accepted emission mechanism for the Earth's auroral kilometric radiation (AKR) for Jupiter's decametric radiation (DAM) and for analogous radiation from the other giant planets. It is also favored for solar microwave spike bursts and for the very bright emission from some flare stars.

An essential requirement for radiation to be observed is that the radiation be able to escape from the source. The only radiation that can escape to infinity is in either the magnetoionic o-mode or x-mode above their respective cutoff frequencies at the plasma frequency, ω_p, and at

$$\omega_x = \frac{1}{2}\omega_B + \frac{1}{2}\left[\omega_B^2 + 4\omega_p^2\right]^{1/2} \approx \omega_B + \omega_p^2/\omega_B\,, \tag{1}$$

where ω_B is the electron cyclotron frequency. The approximation in (1) applies for $\omega_p << 1/2\omega_B$. The polarization of the radiation, which is an important signature of the emission process, results from a preference for one mode over the other.

Plasma emission involves an instability that generates microturbulence in Langmuir (or other) waves that cannot themselves escape. Escaping radiation results from nonlinear interactions in the plasma in which the Langmuir wave, at $\omega_L \sim \omega_p$, $\mathbf{k}_L$ beats with some disturbance at ω', $\mathbf{k}'$ to produce the escaping radiation at

$$\omega = \omega_L + \omega' \ , \ \mathbf{k} = \mathbf{k}_L + \mathbf{k}' \ . \tag{2}$$

Such a three-wave process produces fundamental (F) plasma emission, $\omega \approx \omega_p$, for a low-frequency disturbance, $|\omega'| << \omega_p$, and second harmonic emission (H) when both waves are Langmuir waves, $\omega' = \omega_L$. For type III solar bursts and for plasma emission from planetary bow shocks there is compelling evidence that an electron beam instability generates the Langmuir waves. However, there are other examples of plasma emission, notably type I solar radio emission, where the source of the Langmuir waves remains unclear.

ECME is a direct emission process in which the instability produces emission just above ω_B. It strongly favors the x-mode, which has the electric vector rotating in the same sense as electrons gyrate. In order that the frequency of the growing waves to exceed ω_x, one requires $\omega_B > \omega_p$. This requirement is satisfied in density cavities in the Earth's auroral zones, where the AKR originates, and in Jupiter's magnetosphere in the source region for DAM. In these cases the ECME instability is attributed to electrons precipitating from the magnetosphere into the auroral regions.

The theory of these two familiar cases of radiation from plasma instabilities is reviewed in §2 and a general summary of the applications is presented in §3. The concept of marginal stability and its relevance to the observed distribution functions are discussed in §4. A statistical method, called stochastic growth theory, for treating unstable systems in which the growth occurs sporadically in many localized regions is summarized in §5. Maser theories need to be modified to treat narrowband features in the emission and some ideas relating to phase-coherent mechanisms are outlined in §6.

2 Theory of maser emission processes

The standard theories for both plasma emission and ECME rely on maser emission processes. These are the bump-in-tail instability and the loss-cone driven cyclotron maser instability, respectively. To be effective such instabilities need to be driven, in the sense that the feature in the distribution function that causes the wave growth needs to be continuously regenerated. In maser language, a "pump" is required to drive the maser. In the absence of other limiting effects, such instabilities saturate due to quasilinear relaxation, which is the back reaction to the maser emission. Quasilinear relaxation acts such as to reduce the feature that drives the maser. Saturation may also occur due to various nonlinear effects, which become of increasing importance with increasing energy density in the growing waves.

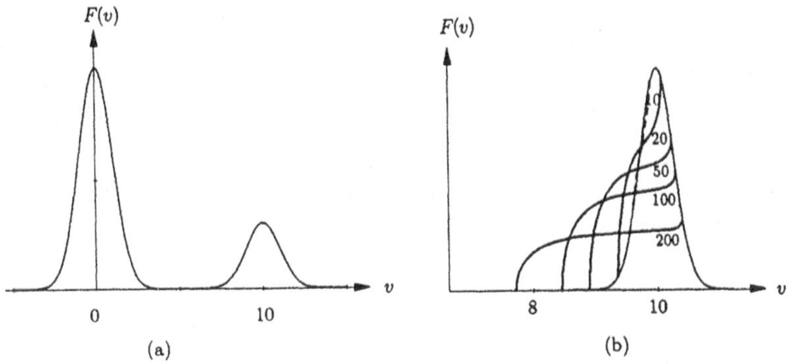

Figure 1: The evolution of the bump-in-tail instability is illustrated (Grognard 1975). (a) The initial distribution causes waves to grow for $v_\phi = v$ in the range where $dF(v)/dv$ is positive. (b) As the growth proceeds, the bump in the tail is eaten away to form a plateau $dF(v)/dv \approx 0$ that extends to lower v as time (denoted by numbers in units of the initial growth time) evolves.

2.1 The bump-in-tail instability

The simplest theory for the bump-in-tail instability involves one-dimensional distributions of streaming electrons and Langmuir waves. Let $F(v)$ be the distribution of electrons integrated over the velocity components perpendicular to the direction of streaming. Let $W(v_\phi)dv_\phi$ be the energy density in Langmuir waves propagating along the streaming direction with phase speed in the range v_ϕ, to $v_\phi + dv_\phi$. A resonant interaction occurs between electrons and Langmuir waves for $v = v_\phi$. Electrons with small $|v - v_\phi|$ tend to be dragged into resonance. The energy gained by electrons with $v \lesssim v_\phi$, comes from the Langmuir waves, and the energy lost by electrons with $v \gtrsim v_\phi$, goes into the Langmuir waves. Maser emission occurs at $v_\phi = v$ for $dF(v)/dv > 0$, in which case there are more electrons with $v \gtrsim v_\phi$ than with $v \lesssim v_\phi$. An idealized example of a bump-in-tail distribution is illustrated in Figure 1.

The growth of the Langmuir waves may be described by

$$\frac{dW(v_\phi)}{dt} = -\Gamma_L(v_\phi)W(v_\phi), \quad \Gamma_L(v_\phi) = -\frac{\pi\omega_p}{n_e} v_\phi^2 \frac{dF(v_\phi)}{dv_\phi}, \tag{3}$$

where Γ_L is the absorption coefficient for the Langmuir waves. Maser emission corresponds to negative absorption. Quasilinear relaxation is described by

$$\frac{dF(v)}{dt} = \frac{\partial}{\partial v} D(v) \frac{\partial F(v)}{\partial v}, \qquad D(v)\frac{\pi\omega_p}{n_e m_e} vW(v). \tag{4}$$

For an *inhomogeneous beam*, which starts from $x = 0$ at $t = 0$, one is to interpret d/dt on the left of (4) as $\partial/\partial t + v\partial/\partial x$. The "pump" is the tendency of $dF(v)/dv$ to increase at the head of the beam due to faster electrons outpacing slower electrons. Quasilinear relaxation opposes this tendency, and the two effects can balance to produce a self-consistent distribution.

One detail of the growth that is important in the following discussion is the characteristic timescale on which quasilinear relaxation occurs:

$$t_{\rm ql} \sim \omega_p^{-1} \frac{n_e}{n_{\rm b}} \frac{W_{\rm b}}{W_L}, \tag{5}$$

where $n_{\rm b}$ and $W_{\rm b}$ are the number density and energy density, respectively, in the beam, and W_L is the energy density in the Langmuir waves. Saturation occurs when W_L becomes comparable with $W_{\rm b}$. Then (5) implies that $t_{\rm ql}$ is several tens times the typical growth time $\Gamma_L^{-1} \sim \omega_p^{-1}\, n_e/n_{\rm b}$.

2.2 The plasma emission processes

Three-wave interactions involving Langmuir waves and ion acoustic waves can lead to plasma emission through the processes outlined in Figure 2. The ion acoustic waves could be present initially, or they could be generated through the three-wave decay process itself. For production of H emission, some of the Langmuir waves generated by the stream must be scattered into the backward (anti-streaming) direction before the coalescence can proceed. The reason is that the initial waves have **k**-vectors approximately along the streaming direction, whereas (2), with the wavenumber of the transverse waves much smaller than that of the Langmuir waves, requires that coalescing Langmuir waves meet approximately head on.

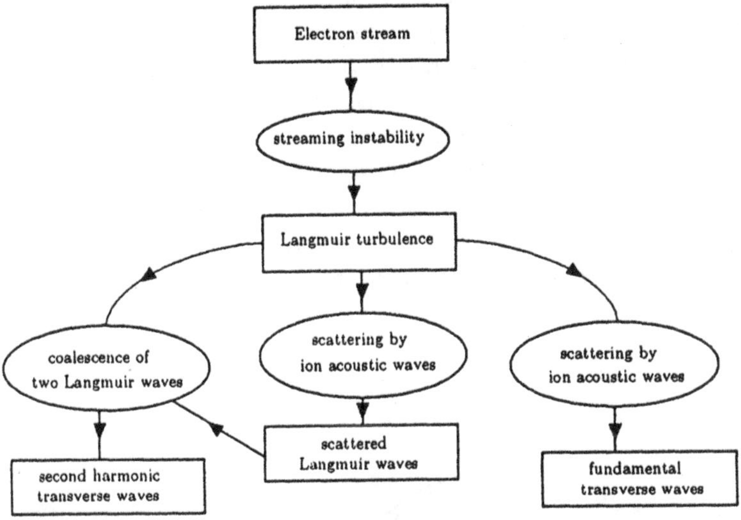

Figure 2: Flow diagram for a variant of the theory of Ginzburg and Zheleznyakov [1958] for the generation of plasma emission. In other variants the processes indicated that involve ion acoustic waves are replaced by other nonlinear plasma processes.

The polarization signature of plasma emission is o-mode emission. F emission should be 100 per cent polarized in the o-mode: the sum of the Langmuir and ion acoustic frequencies, for relevant parameters, is below ω_x so that x-mode emission cannot be generated in the coalescence process. Type III emission is polarized in the o-mode, but never 100 per cent and for some bursts, especially from near the limb, the polarization can be low or undetectable. In contrast, type I emission can be 100 per cent polarized in the o-mode. The lower percentage polarization for type III emission is attributed to depolarization along the ray path from the source to the observer, but there is no entirely satisfactory theory for such depolarization. H emission is expected to be, and is usually observed to be, weakly polarized in the sense of the o-mode.

2.3 ECME

ECME involves negative absorption near the fundamental or harmonics of the cyclotron frequency, $\omega_B = eB/m$. The gyroresonance condition is

$$\omega(1 - \mu\beta \cos\alpha \cos\theta) - s\omega_B/\gamma = 0 \tag{6}$$

where ω and θ, are the wave frequency and wave angle, respectively, μ is the refractive index, $s = 0, \pm 1, \pm 2, \ldots$ is the harmonic number, and $v = \beta c$, α and $\gamma = (1 - \beta^2)^{-1/2}$ are the speed, pitch angle and Lorentz factor, respectively, of the electron. For a given harmonic and a given wave (given s, ω, μ, θ), the resonance condition (6) defines a resonance ellipse in velocity (β, α) space, which determines those electrons that can resonate with the wave at s. Three examples of resonance ellipses are illustrated in Figure 3. The resonance curve labeled (a) is a vertical line at $\beta\cos\alpha = (\omega - s\omega_B)/\mu\omega\cos\theta$, which corresponds to an ellipse with its center far to the left so that the line is actually an arc of a large ellipse. The resonance curve labeled (b) illustrates an intermediate case where the center is closer to the shaded regions.

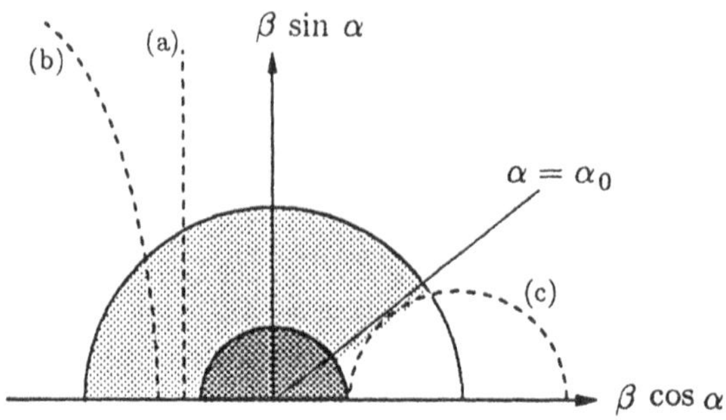

Figure 3: Examples of resonance ellipses in the nonrelativistic region of velocity space. Thermal electrons occupy the darkly shaded region, and nonthermal electrons occupy the lightly shaded region outside the loss cone $\alpha = \alpha_0$. Curves (a), (b) and (c) are discussed in the text.

The resonance curve labeled (c) in Figure 3 is the case of interest here. It corresponds to the 'semirelativistic' approximation $\gamma \approx 1 + \frac{1}{2}\beta^2$ in (6), in which case the ellipse reduces to a circle. It is essential that the relativistic correction, $\frac{1}{2}\beta^2$ be retained even for electrons that can otherwise be treated nonrelativistically [*Twiss* 1958; *Wu and Lee* 1979]. The curve (c) is shown to lie within a loss cone, where the distribution is an increasing function of $p_\perp$, and it is this feature which drives the maser. Also, curve (c) is chosen not to intersect the region where the thermal electrons are located; waves corresponding to resonance ellipses that intersect this region suffer strong thermal gyromagnetic absorption.

2.4 The absorption coefficient

ECME occurs when the cyclotron absorption coefficient is negative. The growth rate for ECME is then identified as the magnitude of this absorption coefficient. An explicit expression for the absorption coefficient is [e.g., *Hewitt, Melrose and Rönnmark* 1982]

$$\Gamma(s, \omega, \theta) = -2\pi \int_0^\infty dpp^2 \int_{-1}^{1} d\cos\alpha\, A(s, \omega, \theta; \beta, \alpha)\Delta f \tag{7}$$

with $p = \gamma\beta mc$ and where $A(s, \omega, \theta; \beta, \alpha)$ involves Bessel functions and a δ-function expressing the resonance condition (6). For an azimuthally symmetric distribution, the integral over azimuthal angle in (7) is trivial, and on carrying out the other integrals using the δ-function, the remaining integral may be rewritten as an integral around the resonance ellipse. The sign of the absorption coefficient is determined by the sign of

$$\Delta f = \left[\frac{s\omega_B}{\omega \sin\alpha}\frac{\partial}{\partial p \sin\alpha} + \frac{\mu p \cos\theta}{mc}\frac{\partial}{\partial p\cos\alpha}\right] f, \tag{8}$$

where $f = f(p, \cos\alpha)$ is the distribution function of the electrons. The two terms in (8) allow the possibility of a *perpendicular-driven maser*, or a *parallel-driven maser*, respectively [e.g., *Melrose* 1986]. The perpendicular-driven maser is favored for ECME.

2.5 Fundamental emission in the x-mode

The most favorable case for ECME is emission at $s = 1$ in the x-mode above the cutoff frequency at ω_x, cf. (1). Notable features of this important case are:

1. Fundamental emission at $\omega \approx \omega_B$ can be in the x-mode only in a plasma with $\omega_p \ll \omega_B$
2. ECME above the cutoff (1) then requires $\omega \approx \omega_B > \omega_p^2/\omega_B$, and hence that the electrons that drive the instability satisfy $\beta\cos\alpha\cos\theta > \omega_p^2/\omega_B^2$.
3. A region around $\theta = \pi/2$ is forbidden for fundamental x-mode ECME.
4. The small range of allowed resonance ellipses, cf. Figure 2.3, for fundamental x-mode ECME imply that the emission is restricted to narrow ranges of frequency and angle. The angular restriction is to a thin surface of a hollow cone with half angle $\theta \approx \cos^{-1}\beta_0$, where $\beta\cos\alpha = \beta_0$ is the center of the resonance ellipse.

2.6 The loss-cone driven ECME

A loss-cone distribution has a deficiency of electrons with small pitch angles, as illustrated in Figure 2. This implies $\partial f/\partial p_\perp > 0$ within the loss cone, allowing perpendicular-driven ECME. Let $\alpha = \alpha_0$ define the edge of the loss cone. The growth rate is given approximately by

$$|\Gamma_x| \approx \frac{\pi\omega_p^2 c^2}{\omega_B}\frac{n_1}{n_e}\frac{1}{\langle v\rangle^2 \sin\alpha_0}, \tag{9}$$

where n_1 and $\langle v\rangle^2$ are the number density and the mean square velocity, respectively, of the electrons that drive the instability.

A favorable pump for ECME is driven precipitation of energetic electrons from a magnetosphere or other closed magnetic structure. Conservation of the magnetic moment, $p_\perp^2/B$, implies that energy in the electron's parallel motion is converted into perpendicular energy as the electron propagates into a region of increasing B. Electrons with small α precipitate, and the reflected electrons then have a loss-cone anisotropy. Once the x-mode waves are generated through ECME, quasilinear relaxation tends to drive electrons from higher to lower $p_\perp$, tending to fill the loss cone.

Saturation of ECME occurs as the free energy becomes exhausted, which occurs after several tens of growth times. In the discussion below the saturation timescale is assumed to be $\sim 30/|\Gamma_x|$

3 Comparison with observations

The foregoing theories are quite successful in accounting for many of the qualitative and semiquantitative properties of the observed plasma emission and ECME. However, there are detailed features of the observations that are not explained in a satisfactory manner by existing theory, notably those that involve short time structures, small spatial variations and narrow frequency structures.

3.1 Solar type III radio bursts

The most successful application of the theory of plasma emission to solar radio bursts is to type III bursts. These are due to electron streams that can propagate through the solar corona into the interplanetary medium (IPM) and out to beyond the orbits of the giant planets.

The most important qualitative predictions of the simple theory of plasma emission are that there should be two and only harmonic bands, from a given source at a given time, and the polarization of F emission should be 100 per cent in th o-mode. Type III bursts in the solar corona roughly exhibit these properties. However, the harmonic ratio is typically less than two, radioheliography shows the F and H emissions at a given time coming from different heights in the corona, and the polarization of F emission is never 100 per cent. Explanations for these properties are sought in terms of propagation effects, and are not interpreted as arguments against the accepted theory.

3.2 Type III bursts in the IPM

Type III bursts in the IPM can be observed in situ. The general features of the theory are confirmed, in that the Langmuir waves appear when the electron beam arrives, and the distribution of electrons has the properties required to cause the Langmuir waves to grow. More specifically, the distribution is consistent with the self-consistent distribution that is found when the one-dimensional quasilinear equation (4) is solved numerically for an inhomogeneous beam [*Grognard* 1984]. Furthermore, there is evidence for ion acoustic waves that appear in association with the Langmuir waves and are consistent with those required by the theory for the three-wave interaction [*Lin et al.*, 1981].

One feature of the observations that is not consistent with simple theory is the spatial distribution of the Langmuir waves, which are highly inhomogeneous. The amplitude of the electric field varies by several orders of magnitude on a typical scale, ~ 100 km, so that the Langmuir waves are observed only in isolated, intense clumps [*Gurnett and Anderson* 1977]. Another feature of the observations is the relative intensity of F to H emission. Although this ratio increases with decreasing frequency for bursts in the solar corona, early investigations suggested that a reversal of this trend, and a changeover to H emission, occurs in the frequency range 10-1 MHz, where few data are available. However, a more careful investigation [*Dulk, Steinberg and Hoang* 1984] showed that type III bursts in the IPM are initially dominated by F emission, with H emission taking over with roughly the same intensity at about the peak in the intensity at fixed frequency.

3.3 Other examples of plasma emission

There are several other contexts in which plasma emission is observed, and which have significantly different plasma and beam parameters from type III bursts. One example is plasma emission from the Earth's bow shock and other planetary foreshocks, where beamed electrons, spiky Langmuir waves and ion acoustic waves are all observed [*Lin et al.* 1981]. The theory of plasma emission for type III bursts explains the broad features of the observed emission in this case [e.g., *Melrose and Cairns* 1985]. Another example is plasma emission in the auroral ionosphere, which is of particular relevance to the discussion below of fine structure on short time scales. Relatively high time-resolution data from high-altitude rocket soundings are available [*Ergun et al.* 1991a], and show large amplitude, localized Langmuir waves associated with localized streaming electron distributions. However, the high particle count rate is achieved at the expense of resolution in velocity space, and expected signatures of electron distribution cannot be resolved [*Ergun et al.* 1991b]. There is indirect evidence from fine structure in auroral arcs that such highly localized Langmuir waves may be a common feature of the auroral ionosphere [e.g., *Borovsky* 1995].

3.4 AKR and DAM

The data on AKR and DAM provide complementary support for the interpretation in terms of ECME. One the one hand, the source region for AKR has been probed directly by spacecraft, providing data on the precipitating electrons, and on the other hand, DAM is unaffected by propagation effects and exhibits the expected features of the angular distribution and of the (elliptical) x-mode polarization.

AKR arises from auroral density cavities, which are the only regions in the terrestrial magnetosphere where the condition $\omega_p \ll \omega_B$ required for ECME is satisfied. AKR correlates with 'inverted V' electron precipitation events [*Gurnett* 1974], and the distribution of inverted V electrons exhibits the loss-cone feature required to drive ECME [*Omidi and Gurnett* 1982; *Melrose, Rönnmark and Hewitt* 1982]. Also, AKR is predominantly in the x-mode at the fundamental $\omega \approx \omega_B$. All these features support the theory for ECME. However, the predicted angular distribution is not clearly in evidence, presumably because of propagation effects, such as reflections from the edges of the density cavities. Similarly, a relatively weak o-mode component may be due to reflections from the edges of the density cavities [*Hayes and Melrose* 1986; cf. *Oya* 1974 for a contrary view].

For DAM, the emission pattern was first inferred by Dulk (1967) to be on the surface of a hollow cone opening angle of $\approx 80°$ and a thickness of $\approx 1°$. This bizarre pattern is predicted by the theory of ECME. The dynamic spectrum measured during fly-bys of Jupiter show nested arc structures which are also consistent with this angular pattern [e.g., *Gurnett and Goertz* 1981]. The polarization of DAM is elliptical [e.g., *Dulk, Leblanc and Lecacheux* 1994], and consistent with x-mode emission, mostly at angles θ close to $\pi/2$, as implied by simple theory [cf. *Willes, Melrose and Robinson* 1994]. It is thought that DAM is due to ECME from ~ 500 keV electrons with the source located on the magnetic field lines that pass through the orbit of Io [e.g., *Goldstein and Goertz* 1983].

Despite the strong evidence in support of the ECME theory, there are features of the data that are not readily explained in terms of the theory. Two are discussed below. First, although the distribution of inverted V electrons in the source region of AKR implies growth for ECME, the growth rate is too small to account for the observed emissions. Second, there are fine structures in DAM and in AKR that seem to require phase-coherent interactions, which cannot be treated using a maser theory.

4 Marginal stability

Both the distribution functions of the electrons that produce type III bursts in the IPM and of the inverted V electrons that produce AKR imply growth rates that are too slow to account for the observed emissions. There is a plausible explanation: the timescale for measurement of the distribution functions is long compared with the timescale for saturation of the instability, so that the observed distribution should be in an average marginally unstable state.

4.1 The marginal stability condition

A balance between the process (pump) that tends to drive a system unstable and the back reaction (quasilinear relaxation) that tends to restore stability leads to a marginally stable distribution [e.g., *Manheimer and Lashmore-Davies* 1989]. The marginally stable state that results for the beam instability is a plateau distribution of particles $dF/dv \approx 0$, and that for the loss-cone instability is a partially filled loss cone with $\partial f/\partial p_\perp \approx 0$. In an idealized situation, the back reaction should cause the growth rate to decrease to an arbitrarily small value. In practice various damping and loss mechanisms need to be overcome before the waves can grow, and there is a large difference in the timescales on which the pump and the back reaction operate. These effects favor the marginal state being approached only in an average sense, with strong localized, sporadic growth rather than weak uniform growth.

4.2 Threshold for effective wave growth

In order for the waves to grow at all in an instability, the (negative) contribution to the absorption coefficient for the waves in question must exceed the effective damping rate, which may be due to Landau damping (Ld), collisional (coll), scattering (scat) out of the region of **k**-space where the waves grow, or escape (esc) from the region or coordinate space where growth is effective. This condition may be written in the form

$$\left|\Gamma_{\text{growth}}\right| > \Gamma_{\text{Ld}} + \Gamma_{\text{coll}} + \Gamma_{\text{scat}} + \Gamma_{\text{esc}} \, . \tag{10}$$

For the growth of Langmuir waves in the IPM the most important term of the right hand side of (10) is that due to scattering. *Muschietti, Goldman* and *Newman* (1985) pointed out that the turbulence observed in the IPM has **k**-values that allow the three-wave beat conditions (2) to be satisfied for the scattering of one Langmuir wave into another. This process causes the Langmuir waves to diffuse in angle. The waves then spend only a fraction of their time in the forward direction, where they tend to grow, and spend a larger fraction of their time in other directions, where they are damped. The growth rate in the forward direction needs to be much larger, than would otherwise be required, in order to lead to net growth.

A similar problem arises with ECME, but for a somewhat reason. The group speed of the growing fundamental x-mode waves is close to c, and hence the waves propagate out of a region of size L at a rate $\Gamma_{\rm esc} \sim c/L$. The size L is restricted by the spatial scale, L_B, of variation in ω_B and the narrow bandwidth, $\Delta\omega$, to which growth is restricted. This implies $\Gamma_{\rm esc} \sim (\Delta\omega)\ (c/L_B)$. Then (10) requires a relatively large intrinsic growth rate for the maser in order to produce significant net growth. An alternative suggestion for AKR is that the growth rate is in fact relatively modest, and that effective growth requires multiple crossings of the growth region due to reflections at the edges of the auroral cavity [*Calvert* 1982].

4.3 Localized regions with strong growth

As already remarked, both for type III events in the IPM and for inverted V electrons, the growth rates estimated for the measured unstable distributions are too modest to account for the observed emissions. However, this need not imply an inconsistency with the data because the relatively large growth rates required imply correspondingly short timescales for quasilinear relaxation. Comparison with the timescales for measurement of the distribution functions shows that one must be measuring the relaxed distributions [e.g., *Melrose* 1989b], which should be in a marginally unstable state according to theory. The slow growth rates inferred for the measured distributions are consistent with this inference. However, it then follows that the features that must drive the instability can be present for too short a time to be measured by the instruments so far available on spacecraft.

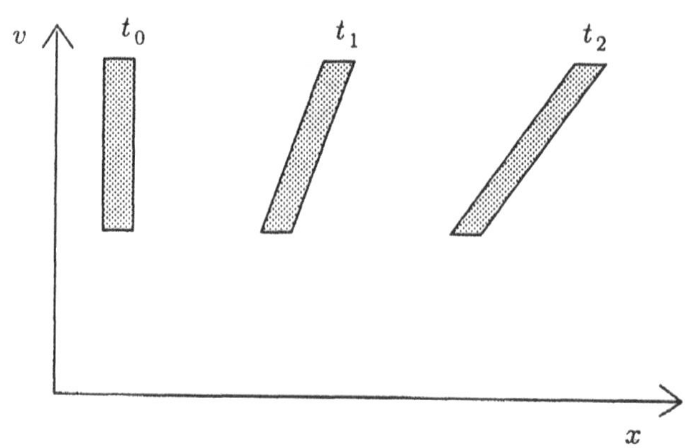

Figure 4: An initial ($t = t_0$) spatial localization of particles is indicated by a shaded region along the v-axis; at later times (t_1, t_2) this region becomes sheared, implying an increasing gradient along the v-axis [Melrose and Goldman 1987].

For type III beams in the IPM and for precipitating electrons in AKR and DAM, the conventional pumps are faster electrons outpacing slower electrons and the increase in $p_\perp$ in a magnetic mirror, respectively. Both operate on timescales of order the electron propagation time across the system of interest. A large growth rate requires a large gradient in velocity space, and such a large average gradient builds up only slowly. Due to the large mismatch between the timescales for build up of the unstable distribution and for its relaxation one expects intense sporadic bursts of growth, rather than slow uniform growth.

Localized regions of enhanced growth require features in the distribution function that are localized in coordinate space with large gradients in velocity space. In the absence of data on a sufficiently short timescale to resolve such features that drive the instability, one can only speculate on the processes involved in setting them up. One idea is that a distortion can arise in velocity space from a localized distortion in coordinate space, as illustrated in Figure 4. Suppose a burst of wave growth causes a pile up in coordinate space due to the slow spatial diffusion through the region of enhanced wave turbulence. The initial ($t = t_0$) distribution function, $f_0(x, v) = f(x, v, t_0)$, then has a maximum as a function of x, at $\partial f(x, v, t_0)/\partial x = 0$. At later time the distribution is of the form $f(x, v, t) = f_0(x - v(t - t_0), v)$, so that one has $\partial f/\partial v = \partial f_0/\partial v - t\partial f_0/\partial x$. For a plateau distribution, $\partial f_0/\partial v \approx 0$, as an initial spatial gradient evolves it produces an increasingly sharp v-space gradient. After a sufficient time has elapsed, a local instability tends to develop. The resulting enhanced level of waves tends to cause a localized pile up in coordinate space, so that the process can then repeat itself.

5 Stochastic growth theory

The foregoing arguments suggest that systems close to marginal instability should exhibit bursts of localized wave growth. A general method for treating such systems statistically is in terms of stochastic growth theory [e.g., *Robinson* 1995]. Although quasilinear relaxation and the plasma emission processes are affected locally by the highly inhomogeneous distribution of waves, the average effects are not particularly sensitive to the spatial distribution of the waves.

5.1 Stochastic growth

The basic idea in stochastic growth theory is that the growth rate, $-\Gamma(t)$, is a stochastic variable. The wave intensity, $W(t)$, varies according to $dW(t)/dt = -\Gamma(t)W(t)$. On integrating this equation, it is convenient to introduce a gain factor, $G(t)$, which satisfies $dG(t)/dt = -\Gamma(t)$. Then one has $W(t) = \exp[G(t)]$. The problem is then amenable to a statistical description based on the probability distribution for values of G and Γ [*Robinson* 1995].

In the simplest case, Gaussian statistics are assumed, so that all higher order correlations of $G(t)$ can be expressed in terms of the average, $\langle G\rangle$, and the variance, σ_G^2, of G. The probability distribution of a value G for the growth is then

$$P(G) = \frac{1}{(2\pi)^{1/2}\sigma_G} \exp\left[-\frac{(G-\langle G\rangle)^2}{2\sigma_G^2}\right]. \tag{11}$$

The statistics of the electric field fluctuations follow from $G = 2\ln(E/E_0)$, where E_0 is some characteristic initial field. The probability distribution for the electric field fluctuations is

$$P(E) = \frac{1}{E(2\pi)^{1/2}\sigma_G} \exp\left[-\frac{[\ln(E/E_0)-\langle G\rangle]^2}{2\sigma_G^2}\right]. \tag{12}$$

This provides a model to compare with the observed distributions of electric fields in the clumpy Langmuir waves.

5.2 Application to type III bursts

For type III bursts in the IPM, the average growth rate is ~ 0.1 s^{-1} and the scattering rate is of order ~ 10 s^{-1} [e.g., *Melrose and Goldman* 1987; *Robinson* 1995]. Applying stochastic growth theory to this case, *Robinson* [1995] estimated that the ratio of the standard deviation to the mean of the growth rate is ~ 4. These parameters were used to predict the statistical distribution of electric fields in the Langmuir waves using (12), and the comparison provides general support for the model [*Robinson* 1992, 1995; *Robinson, Cairns and Gurnett* 1993]. The agreement is encouraging, suggesting that statistical fluctuations in the growth rate can indeed account for the statistical properties of the inhomogeneous distribution of Langmuir waves in a simple way.

5.3 Plasma emission for clumpy Langmuir waves

The highly inhomogeneous distribution of Langmuir waves in type III events modifies both the quasilinear evolution of the particles and the plasma emission processes, compared with a homogeneous distribution of waves with the same average energy density, $\overline{W}_L$. However, the effect is quite minor for the distribution of particles. Indeed, the quasilinear relaxation resulting from encounters with a statistical distribution of clumps of waves, occurs at the same rate as for the equivalent homogeneous distribution [*Melrose and Cramer* 1989]. The effect of inhomogeneous wave distribution on the plasma emission processes is more pronounced. However, these processes tend to saturate when the brightness temperature of the radiation approaches that of the Langmuir waves, and there is some observational support for a model in which this saturation level is approached [*Melrose* 1989a]. The inhomogeneous distribution favors H emission, which develops at a rate $\propto W_L^2$, rather than F emission, which develops at a rate $\propto W_L$, compared with a homogeneous distribution with the same $\overline{W}_L$. However, this effect is partly offset by localization of the F component due to multiple scatterings from local density inhomogeneities [*Melrose* 1989b]. If both harmonics reach saturation, then the resulting emission is not sensitive to the details of the spatial distribution of Langmuir waves. Alternatively, a detailed investigation of a model for plasma emission due to a clumpy distribution of Langmuir waves [*Robinson, Willes and Cairns* 1993] suggests that saturation of the beam instability is due to the onset of the decay instability $L \rightarrow L' + S$. The model appears capable of accounting for the observed properties of type III events in the IPM, including the changeover from F to H emission, without appealing to the saturation of the plasma emission processes.

6 Fine structures in observed emission

There are fine structures in the spectra of sources attributed to ECME that cannot be explained within the context of the simple maser theory. One possible explanations is by analogy with discrete VLF emission.

6.1 Observations of fine structure

Ellis [1973, 1974] reported fine structures in DAM at $\lesssim 10$ MHz. These structures drift rapidly in frequency, ~ 10 MHz s^{-1}, predominantly from higher to lower frequencies. The bandwidth is very narrow, with some bursts unresolved by the instrumental bandwidth of 2 kHz. The qualitative properties of these fine structures are analogous to those of discrete VLF emissions from the terrestrial magnetosphere: including triggering and interactions, such as abrupt changes in ω-t slope, when burst spectra cross constant-frequency emissions. Similar fine structures are also present in AKR [e.g., *Gurnett and Anderson* 1981; *Morioka, Oya and Miyatake* 1981]. As with the fine structures in the Jovian bursts, there is a close similarity with discrete VLF emissions.

6.2 Possible analogy with discrete VLF emission

The narrow frequency spectra and changing frequency as a function of time suggest that phase coherent processes must be involved in these fine structures. *Melrose* [1986] proposed a phase-bunching model based on the phenomenological model of Helliwell [1967] for discrete VLF emissions. In Helliwell's model the source region is identified as a localized interaction region (IR) in which the relevant electrons flow in on one side of the IR where they are phase-bunched by an outgoing wave, and these electrons give up energy to an incoming wave as they leave the opposite side of the IR. The IR is moving, and its motion is such that the resonant electrons remain in phase with the waves for an exceptionally long time. This requirement is expressed by the condition, cf. (6),

$$d\left[\omega(1-\mu\beta\cos\alpha\cos\theta)-s\omega_B/\gamma\right] = 0 \ , \tag{13}$$

where the differential relates the change in frequency with the spatial changes in the other variables along the path of the IR. *Melrose* [1986] argued that for (13) to be satisfied for fundamental x-mode emission, the gradient $d\omega_B/dz$ due to the changing magnetic field strength must be balanced by an acceleration due to a parallel electric field, $E_{\|} = -\,d\Phi/dz$. Then (13) is to be evaluated subject to conservation of energy, $d(m_e c^2\gamma - e\Phi) = 0$, and conservation of the adiabatic invariant, $d(\gamma^2 v_\perp^2/\omega_B) = 0$. The model then require $E_{\|} \sim E_0 = -(m_e c^2/2e\omega_B)(d\omega_B/dz)$.

The phase-coherent interaction saturates when the trapping frequency,

$$\omega_t = \left(\frac{eE_x v_\perp \omega_B}{m_e c^2}\right)^{1/2}, \tag{14}$$

where E_x is the electric field of the x-mode waves, is of order the inverse of the resonance time. The resonance time is estimated as $\sim L/|v_{\|} - v_{IR}|$, which is the time an electron remains in the IR, of length L, moving along the magnetic field lines with velocity v_{IR}. The bandwidth of the observed fine structures should be of order this time, that is, $\sim \omega_t$.

The phenomenological model of *Helliwell* [1967] for discrete VLF emissions requires a source of free energy, and this is not included explicitly in the model. Indeed an unsatisfactory feature of the models is that it does not link up naturally to more conventional plasma theory [e.g., the review by *Matsumoto* 1979]. The accepted source of free energy is the pitch angle anisotropy which is in a marginally stable state, with losses due to scattering of into the loss cone by whistlers balancing inward diffusion of energetic electrons [e.g., *Kennel and Petschek* 1966]. In modifying this model to x-mode waves, *Melrose* [1986] argued similarly that the electron distribution that leads to AKR in the marginal state for ECME. Phase bunching can then trigger the release of available free energy, as discussed by *Winglee* [1985]. Evidence that AKR can be triggered by an external signal [*Calvert* 1985], which is a characteristic feature of discrete VLF emission, is further evidence for the role of phase bunching.

7 Discussion and conclusions

Although theories of plasma emission and ECME were developed in detail the early 1970s and early 1980s, respectively, important aspects still remain uncertain. One uncertainty concerns how the instabilities saturate. More important is the realization that the instabilities develop sporadically in highly localized regions, rather than smoothly throughout the source region. As a consequence in those cases where the distribution function can be measured, the measured distribution is the marginally unstable average distribution. Much higher count rates are required to identify the features that actually cause the instabilities. The sporadic, localized waves growth may be described using statistical methods, with stochastic growth theory being an example. However, further development of the relation between microscopic and macroscopic theories is needed [e.g., *Burch et al.* 1995].

References

Borovsky, J.E., 1995, Fine-structure in auroral arcs: an unexplained phenomenon, in *Space Plasmas: Covpling Betvteen Small and Medium Scale Processes,* M. Ashour-Abdalla, T. Chang & P. Dusenbery (eds), American Geophysical Union, Washington DC, pp. 255-267.

Burch, J.L., Lin, C.S., Menietti, J.D., & Winglee, R.M., 1995, Coupling between mesoscale and microscale processes in the cusps and auroral plasmas, *in Space Plasmas: Coupling Between Small and Medium Scale Processes,* M. Ashour-Abdalla, T. Chang & P. Dusenbery (eds), American Geophysical Union, Washington DC, pp. 269-283.

Cairns, I.H., & Melrose, D.B., 1985, A theory for $2f_p$ radiation upstream of the Earth's bowshock, *J. Geophys. Res. 90*, 6637-6640.

Calvert, W., 1982, A feedback model for the source of auroral kilometric radiation, *J. Geophys. Res. 87*, 8199-8214.

Calvert, W., 1985 Auroral kilometric radiation triggered by type II solar radio bursts *Geophys. Res. Lett. 12*, 377-380

Dulk, G.A., 1967, Apparent changes in the rotation rate of Jupiter, *Icarus* 7, 173-182.

Dulk, G.A., Leblanc, Y., & Lecacheux, A., 1994, The complete polarization state of Io-related radio storms from Jupiter: a statistical study, Astron. *Astrophys. 286*, 683-700.

Dulk, G.A., Steinberg, J.-L., & Hoang, S., 1984, Type III bursts in interplanetary space. Fundamental or harmonic?, *Astron. Astrophys. 141*, 30-38.

Ellis, G.R.A., 1973, Hyperfine structure in the radio spectra of Jupiter, *Proc. Astron. Soc. Australia 2*, 191-193.

Ellis, G.R.A., 1974, The Jupiter radio bursts, *Proc. Astron. Soc. Australia 2*, 236 243.

Ergun, R.E., Carlson, C.W., McFadden, J.P., Clemmons, J.H., & Boehm, M.H., 1991a, Langmuir wave growth and electron bunching: results from a wave-particle correlator, *J. Geophys. Res. 96*, 225-238.

Ergun, R.E., Carlson, C.W., McFadden, J.P., Tonthat, D.M., Clemmons, J.H., & Boehm, M.H., 1991b, Observations of electron bunching during Landau growth and damping, *J. Geophys. Res. 96*, 11,371-11,378.

Ginzburg, V.L., & Zheleznvakov, V.V., 1958, On the possible mechanism of sporadic solar radio emission (radiation in an isotropic plasma), *Soviet Astron. AJ 2*, 653-667.

Goldstein, M.L., & Goertz, C.K., 1983, Theories of radio emissions and plasma waves, in A.J. Dessler (ed.) *Physics of the Jovian Magnetosphere,* Cambridge University Press, pp. 317-352.

Grognard, R.J.-M., 1975, Deficiencies of the asymptotic solutions commonly found in the quasilinear relaxation theory, *Aust. J. Phys. 28,* 731-753.

Grognard, R.J.-M., 1984, Partial reconstruction of the initial conditions for streams of energetic electrons associated with a type III burst, *Solar Phys. 94*, 165-170.

Gurnett, D.A., 1974, The Earth as a radio source. Terrestrial kilometric radiation, *J. Geophys. Res. 79*, 4277-4238.

Gurnett, D.A., & Anderson, R.R., 1977, Plasma wave electric fields in the solar wind: Initial results from Helios 1, *J. Geophys. Res. 82*, 632-650.

Gurnett, D.A., & Anderson, R.R., 1981, The kilometric radio emission spectrum: relation to auroral acceleration processes, in S.-I. Akasofu & J.R. Kan *Physics of Auroral Arc Formation*, American Geophysical Union, Washington DC, pp. 341-350.

Gurnett, D.A., & Goertz C.K., 1981, Multiple Alfven wave reflections excited by Io: origin of the Jovian decametric arcs, *J. Geophys. Res. 86*, 717-722.

Hayes, L.M., & Melrose, D.B., 1986, Generation of ordinary mode auroral kilometric radiation from extraordinary mode waves, *J. Geophys. Res. 91*, 221-217.

Helliwell, R.A., 1967, A theory of discrete VLF emissions from the magnetosphere, *J. Geophys. Res. 72*, 4773-4790.

Hewitt, R.G., Melrose, D.B., & Ronnmark, K.G., 1982, The loss-cone driven electron-cyclotron maser, *Aust. J. Phys. 35*, 447-471.

Kennel, C.F., & Petschek, H.E., 1966, Limit on stably trapped particle fluxes, *J. Geophys. Res. 71*, 1-28.

Lin, R.P., Potter, D.W., Gurnett, D.A., & Scarf, F.L., 1981, Energetic electrons and plasma waves associated with a solar type III radio burst, *Astrophys. J. 251*, 364-373.

Manheimer, W.M., & Lashmore-Davies, C.N., 1989, *MHD and Microinstabilities in Confined Plasmas*, Adam Hilger, Bristol, pp. 241-250.

Matsumoto, H., 1979, Nonlinear whistler-mode interaction and triggered emissions in the magnetosphere: a review, in P.J. Palmadesso & K. Papadopoulos (eds) *Wave Instabilities in Space Plasmas*, D. Reidel, Dordrecht, pp. 163-190.

Melrose, D.B., 1976, An interpretation of Jupiter's decametric radiation and the terrestrial kilometric radiation as direct amplified gyroemission, *Astrophys. J. 207*, 651-662.

Melrose, D.B., 1986, A phase-bunching mechanism for fine structures in the auroral kilometric radiation and the Jovian decametric radiation, *J. Geophys. Res. 91*, 7970-7980.

Melrose, D.B., 1989a, The brightness temperatures of solar type III bursts, *Solar Phys. 120*, 369-381.

Melrose, D.B., 1989b, Wave localization in solar type III events, *Solar Phys. 119*, 385-398.

Melrose, D. B., & Cramer, N.F., 1989, Quasilinear relaxation of electrons interacting with an inhomogeneous distribution of Langmuir waves, *Solar Phys. 123*, 343-365.

Melrose, D.B., Dulk, G.A., & Cairns, I.H., 1986, Clumpy Langmuir waves in Type III solar radio bursts, *Astron. Astrophys. 163*, 229-238.

Melrose, D.B., & Goldman, M.V., 1987, Microstructures in type III events in the solar wind, *Solar Phys. 107*, 329-350.

Melrose, D.B., Rönnmark, K.G., & Hewitt, R.G., 1982, Terrestrial kilometric radiation: the cyclotron theory, *J. Geophys. Res. 87*, 5140-5150.

Morioka, A., Oya, H., & Miyatake, S., 1981, Terrestrial kilometric radiation observed by satellite JIKIKEN (EXOS-B), *J. Geomag. Geoelectr. 33*, 37-62.

Muschietti, L., Goldman, M.V., & Newman, D., 1985, Quenching of the beam-plasma instability by large-scale density fluctuations in 3 dimensions, *Solar Phys. 96*, 181-198.

Omidi, N., & Gurnett, D.A., 1982, Growth rate calculations of auroral kilometric radiation using the relativistic resonance condition, *J. Geophys. Res. 87*, 2377-2383.

Oya, H., 1974, Origin of Jovian decameter wave emissions - Conversion from the electron cyclotron plasma wave to the ordinary mode electromagnetic wave, *Planet. Space Sci. 22*, 687-708.

Robinson, P.A., 1992, Clumpy Langmuir waves in type III radio sources, *Solar Phys. 139*, 147-163.

Robinson, P.A., 1995, Stochastic growth theory, *Phys. Plasmas 2*, 1466-1479.

Robinson, P.A., Cairns, I.H. & Gurnett, D.A., 1993, Clumpy Langmuil waves in type III radio sources: Comparison of stochastic-growth theory with observations, *Astrophys. J. 407*, 790-800.

Robinson, P.A., Cairns, I.H. & Willes, A.J., 1994, Dynamics and efficiency of type III solar radio emission, *Astrophys. J. 422*, 870-882.

Robinson, P.A., Willes, A.J., & Cairns, I.H., 1993, Dynamics of Langmuir and ion-sound waves in type III solar radio sources, *Astrophys. J. 408*, 720-734.

Twiss, R.Q., 1958, Radiation transfer and the possibility of negative absorption in radio astronomy, *Aust. J. Phys. 11*, 564-579.

Willes, A.J., Melrose, D.B., & Robinson, P.A., 1994, Elliptically polarized Jovian decametric radiation: An investigation of the electron cyclotron maser mechanism, *J. Geophys. Res. 99*, 21, 203-21, 211.

Winglee, R.M., 1985, Enhanced growth of whistlers due to bunching of untrapped electrons, *J. Geophys. Res. 90*, 5141-5152.

Wu, C.S., & Lee, L.C., 1979, A theory of the terrestrial kilometric radiation, *Astrophys. J. 230*, 621-626.

COSMIC MASERS – A POWERFUL TOOL FOR ASTROPHYSICS

James M. Moran

1 Introduction

Cosmic masers were discovered accidentally in 1965 by radio astronomers who were trying to map the distribution of gas in molecular clouds by measuring the absorption due to OH molecules against thermal background sources [*Weaver et al.*, 1965, *Gundermann*, 1965]. The compelling argument for maser emission was that, although the spectral lines had linewidths appropriate for gas at ~50 K, their equivalent blackbody brightness temperatures were very much higher. Indeed, subsequent interferometric measurements have shown that the brightness temperatures reach as high as 10^{15} K. Figure 1 shows a spectrum of one the earliest known masers, which displays many of their unusual characteristics. Masers occur in molecular clouds that are predominantly made up of molecular hydrogen with trace amounts of other molecules. The most powerful and widespread masers originate from transitions of OH, H_2O, SiO, CH_3OH and NH_3 molecules.

Since 1965 research on masers has flourished. This review focuses on the uses of masers as astrophysical tools. For this purpose the most important characteristic of cosmic masers is their intense point-like emission within narrow spectral lines, which allows for precise measurements of their relative angular positions, velocities, and spectral line shapes. Four topics of current interest are discussed here: (1) masers as tracers of the dynamical properties of star forming regions, supernova remnants, envelopes of late-type stars, and accretion disks in active galactic nuclei; (2) masers as tracers of magnetic fields; (3) masers as distance indicators; and (4) masers as probes of interstellar scattering.

Comprehensive discussions of masers can be found in the book by *Elitzur* [1992a] and in various review articles (e.g., [*Cohen*, 1989], [*Elitzur*, 1992b], [*Reid* and *Moran*, 1988], and [*Watson*, 1994]). The proceedings of a conference devoted solely to cosmic masers provides a broad overview of the field [*Clegg* and *Nedoluha*, 1993]. In addition, the conference proceedings *The Structure and Content of Molecular Clouds, 25 Years of Molecular Radioastronmomy* [*Wilson* and *Johnston*, 1994] contains many articles on masers from an historical perspective.

Cosmic masers and their artificial terrestrial counterparts both rely on population inversion to achieve high brightness, but otherwise they have few similarities. Cosmic masers are one-pass amplifiers, since there is no mechanism for multiple reflections in molecular clouds, and enormous path lengths are required to develop the requisite gain. Hence, their spatial coherence is probably very small, and the beam angles of the emergent radiation are probably on the order of a degree or more. The linewidths are approximately equal to the thermal Doppler widths, but usually somewhat narower. Cosmic masers fluctuate in amplitude on time scales of minutes to years. However, the electric fields appear to be Gaussian random processes and show no evidence of any pulsing phenomena [*Evans et al.*, 1972; *Moran*, 1981].

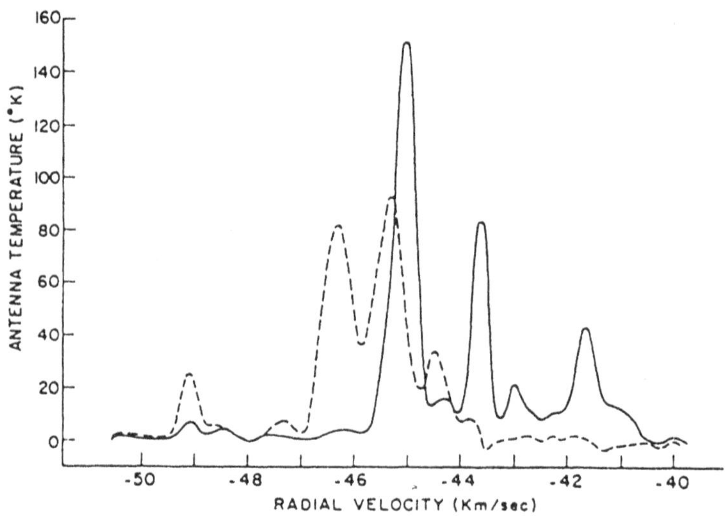

Figure 1: The spectrum of the 1665 MHz transition of OH towards W3(OH). The solid and dashed lines denote right and left-handed circular polarization, respectively [Barrett and Rogers, 1966]. The multiple spectral features, high polarization, narrow linewidths, and high brightness temperatures are evidence for a maser process. Each spectral feature arises from a separate parcel of gas that appears as a maser spot (see Figure 5).

2 Theoretical properties

The observable quantity for any radio source is its flux density (10^{-26} Wm^{-2} Hz^{-1} ≡ 1 Jansky), which is given by

$$F = I\,\Omega_s \qquad (1)$$

where I is the intensity and Ω_s is the solid angle subtended by the source as viewed from the Earth. The intensity can be related to the brightness temperature, T_b, of an equivalent black body through the Rayleigh-Jeans approximation by the relation

$$T_b = \frac{\lambda^2 F}{2k\Omega_s}, \qquad (2)$$

where λ is the wavelength, and k is Boltzmann's constant. The brightness temperatures of masers are so high that no thermal mechanism can plausibly explain the emission. Such high brightness temperatures are possible through the process of population inversion, which requires special circumstances that violate the conditions of local thermodynamic equilibrium. A generic population diagram of a three level maser is shown in Figure 2.

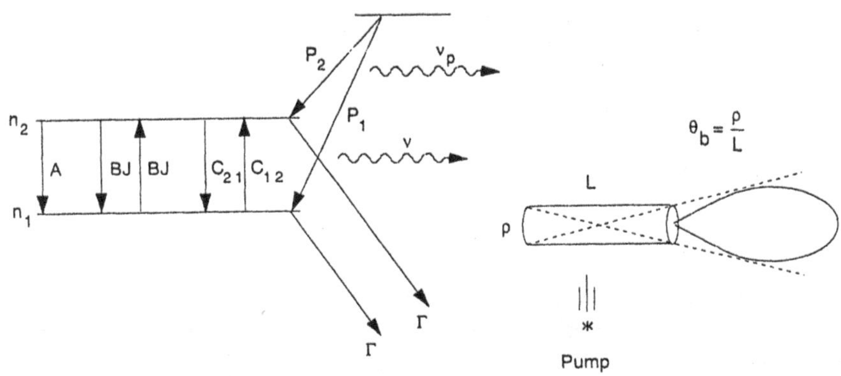

Figure 2: (left) The energy level diagram of a three level maser. (right) Schematic representation of a filamentary maser, and beaming of radiation that results from this geometry.

The population in the upper level of the maser transition can exceed that of the lower level if the pump rate (per molecule) P_2 exceeds the pump rate P_1. Solution to the steady state rate equations, based on the transition processes shown in Figure 2, gives the relation for the population inversion,

$$\Delta n = \frac{\Delta n_0}{1 + 2BJ/\Gamma}, \qquad (3)$$

where, B is the Einstein B coefficient, Γ is the decay rate from the maser levels, J is the angle–average intensity, $J = I\Omega_b/4\pi$, where Ω_b is the angle into which the maser emission is beamed, and Δn_0 is the population inversion that would exist in the absence of a maser radiation field, $\Delta n_0 = n_{12}\,\Delta P/P$, where $\Delta P = P_2 - P_1$ (the differential pump rate per molecule) $P = P_1 + P_2$, and $n_{12} = n_1 + n_2$. The effect of collisions across the maser levels (rates C_{12} and C_{21}) have been neglected). The pump energy for cosmic masers in the Galaxy most likely comes from radiation from a nearby star in the form of infrared radiation or from thermal energy that heats the masing gas through collisions, usually involving shock waves. For population inversion to occur, the system must have reservoirs or heat sinks characterized by two different temperatures. One common way to achieve this situation is for the "used" infrared photons involved in the pump cycle to escape from the maser cloud.

With population inversion, the radiation field can increase exponentially if the pumping action is sufficiently strong, that is, if $\Delta n \simeq \Delta n_0$ (see Eq. 3). The characteristic gain length for exponential amplification is given by

$$L_0 = \frac{8\pi\Delta\nu\Gamma}{A\lambda^2\,\Delta n_0}\,, \tag{4}$$

where A is the Einstein A coefficient, $2h\nu^3B/c^3$, and $\Delta\nu$ is the linewidth. When the radiation field builds up to the point where $2BJ/\Gamma > 1$ (see Eq. 3), it will reduce the inversion and limit the growth of the radiation field (i.e., the maser saturates). In a one–dimensional saturated maser the radiation only increases linearly with the amplification path length. The saturation condition can be written in terms of a brightness temperature as

$$T_s = \frac{2\pi}{k}\,\frac{h\nu}{\Omega_b}\,\frac{\Gamma}{A}\,. \tag{5}$$

where h is Planck's constant and ν is the maser frequency.

A simple conceptual model of a cosmic maser is that of a filamentary structure of cross sectional diameter ρ and length L (see Figure 2). The filamentmay not necessarily be a physical entity; it is defined as the volume overwhich the line–of–sight velocity variations due to turbulence or organized motions are less than the thermal linewidth. In this case the beam angle of the maser is given by

$$\Omega_b \simeq \frac{\pi}{4}\left(\frac{\rho}{L}\right)^2 \tag{6}$$

One of the principal frustrations of modelling the physical properties of masers is that Ω_b cannot be directly observed. Long filamentary tubes are thought to be the common maser geometry. In a limiting form where the middle of the tube is missing, most of its essential properties remain, but the geometry looks more like one maser cloud amplifying another. Chance amplification of a background maser cloud by a foreground cloud at the same line–of–sight velocity may be an important phenomenon in astrophysical masers [*Deguchi and Watson*, 1989]. Filamentary or slab structures may arise naturally in shocked regions [*Elitzur, Hollenbach*, and *McKee*, 1989].

A maser that is saturated is more efficient than one that is unsaturated in terms of pump power: in a saturated maser one pump photon can produce one maser photon. The luminosity, $\mathfrak{L}$, of such a maser is

$$\mathfrak{L} = h\nu\Delta n_0 PV \quad , \tag{7}$$

where V is the volume of the masing gas. From this expression the observed flux density can be written as

$$F = \frac{h\nu P\Delta n_0 L^3}{D^2\Delta\nu} \tag{8}$$

where D is the distance to the maser. The flux density increases as L^3 due to the combined action of linear growth and beam narrowing (Eq. 6). The rather general thermodynamic constraint that each maser photon must be accompanied by an infrared pump cycle photon that escapes from the maser boundaries, leads to a condition known as the *thermodynamic limit* (which can be derived with the use of Eqs. 1, 2 and 8) [see *Goldreich* and *Kwan*, 1974; *Watson*, 1994] and gives the relation

$$T_B = 4\pi T_p\left(\frac{\nu_p}{\nu}\right)^2\left(\frac{L}{\rho}\right)^3 , \tag{9}$$

where T_p and ν_p are the temperature and frequency of the pump.

A typical strong H_2O maser from the 6_{16} - 5_{23} transition at 22 GHz (1.35 cm) in a star–formation region might have the observable quantities: $F = 10^4$ Jy, $\theta_s = 1$ mas, ($\Omega_S = 1.8 \times 10^{-17}$ steradians) $D = 1$ kpc and linewidth 1 kms^{-1}, which at 22 GHz corresponds to $\Delta\nu = 74$ kHz. The linear cross section of the maser, ρ, is 0.7×10^{13} cm and the brightness temperature (Eq. 2) is 1.4×10^{14} K. If T_p and ν_p are assumed to be 400K and 3×10^{13} Hz (100 μm), respectively, Eq. 9 gives $(L/\rho) \simeq 100$, and thus $L \simeq 10^{15}$ cm. Since A = 1.9×10^{-9} s^{-1}, $T_s \simeq 10^{13}$K, and $P = 1$ s^{-1} (the value of the Einstein A coefficient for rotational levels connecting the maser levels) the maser requires $P\Delta n_0 \simeq 30$ (see Eq. 8). With a one percent population inversion (a reasonable value on theoretical models), one percent maser level population, and 5×10^{-5} [H_2O/H_2] ratio (the cosmic abundance ratio), the required H_2 density in the maser cloud is 10^{10} cm^{-3}. Note that it is not possible to increase the maser luminosity arbitrarily, because the gas density cannot exceed the critical value for thermal quenching from the effects of collisions (about 10^{10} cm^{-3} for H_2O masers), and the pump rate per molecule cannot exceed Γ (which for H_2O masers is about 1 s^{-1}).

An interesting new discovery is the large number of H_2O masers among themillimeter and submillimeter wavelength transitions, as shown in Figure 3. These bright lines may be useful for calibrating millimeter and submillimeter wavelength interferometers.

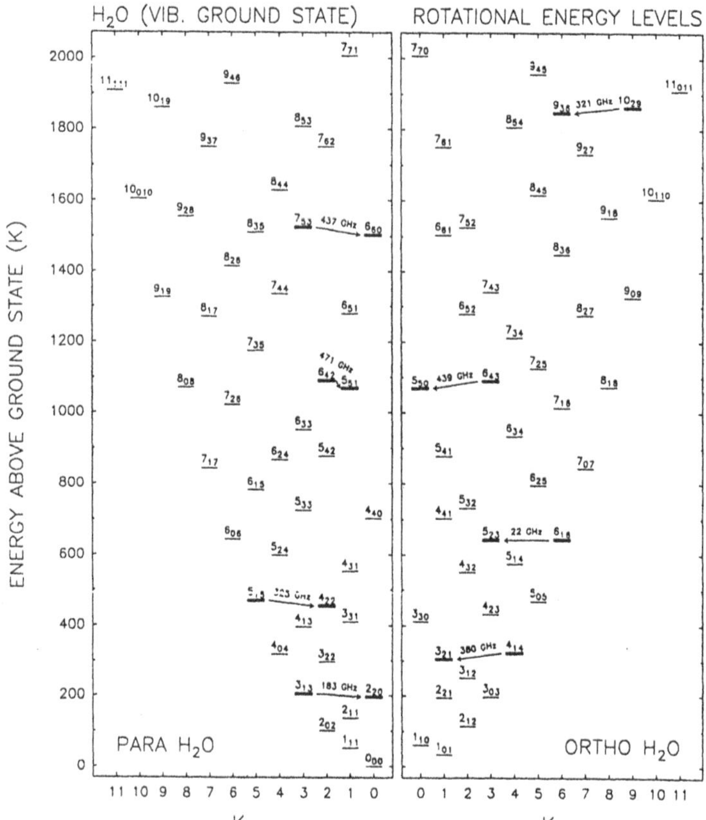

Figure 3: Rotation energy levels of the H_2O molecule up 1400 cm^{-1} above the ground state. The arrows mark the transitions for which cosmic masers have been identified [Menten, 1994]. Additional masers have bee detected in excited vibrational states. The most well studied H_2O maser is the one formed in the 6_{16}-5_{23} transition at 22 GHz (1.35 cm).

3 Dynamical tracers

Masers have proven to be remarkably effective tools for probing the motions of gas around very compact objects, which are not accessible by any other means. Most of this work is possible only through the high angular resolution provided by very long baseline interferometry (VLBI). The newest instrument of this type, the Very Long Baseline Array (VLBA) [*Napier et al.*, 1994], has just begun operation and is expected to have a major impact on maser research. A diagram of the VLBA is shown in Figure 4. The resolution of the VLBA (θ_R = λ/baseline length) is about 300 μas at 1.3 cm wavelength. The relative positions of maser spots in a cluster can be measured to an accuracy of about $\delta\theta \simeq \theta_R/(2SNR)$ where *SNR* is the signal–to–noise ratio. The angular motion of a maser spot in the plane of the sky is

$$\dot{\theta} = 206\left(\frac{V_T}{\text{kms}^{-1}}\right)\left(\frac{D}{\text{kpc}}\right)^{-1} \mu\text{as yr}^{-1}, \tag{10}$$

where V_T is the transverse linear velocity. Hence, the minimum transverse velocity that can be detected in time *T* is

$$V_T \simeq 0.7\left(\frac{D}{\text{kpc}}\right)\left(\frac{T}{\text{kpc}}\right)^{-1} (SNR)^{-1} \text{ kms}^{-1} . \tag{11}$$

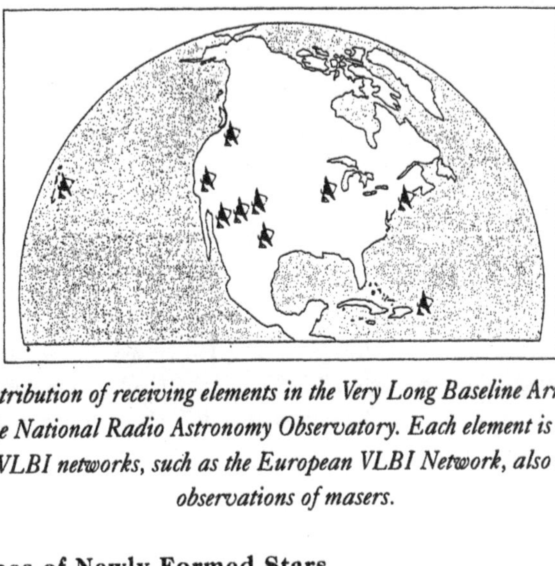

Figure 4: The distribution of receiving elements in the Very Long Baseline Array (VLBA), which is operated by the National Radio Astronomy Observatory. Each element is a 25-m parabolic antenna. Other VLBI networks, such as the European VLBI Network, also routinely schedule observations of masers.

3.1 Envelopes of Newly Formed Stars

Newly formed stars develop winds of outflowing gas and dust, which disperse the material from which they formed [*Shu, Adams,* and *Lizano*, 1987]. These envelopes frequently show maser emission. For example, in W3(OH) the OH molecules in the molecular envelope show that it is probably expanding ahead of the ionization shock front at a velocity of a few kms^{-1}. Figure 5 shows the distribution of these maser spots and their proper motions. Each spot originates from a clump of gaz with a distinct velocity corresponding to a spectral feature, as shown in Fig. 1.

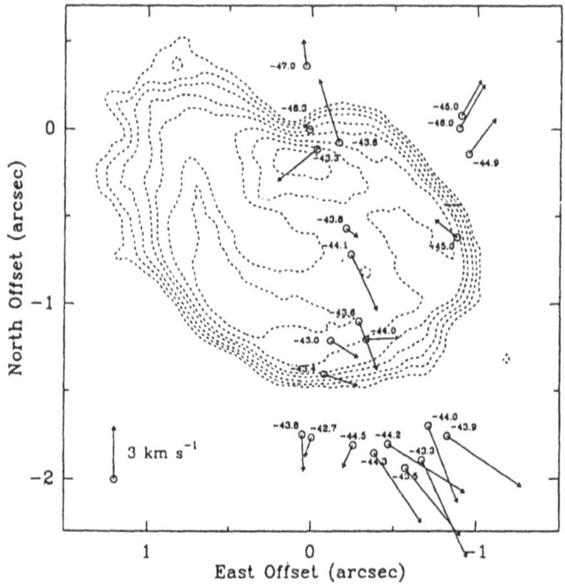

Figure 5: Map of the distributions of OH masers in the source W3(OH). The spots are labelled by their line–of–sight velocities. The spots correspond to the spectral features shown in Figure 1. The transverse velocities are indicated by the arrows (for a distance of 2.2 kpc). The dotted contours denote thermal continuum emission from gas ionized by an embedded O–type star [Bloemhof et al., 1992].

In a nearby region, probably at an earlier stage of development than the OH masers in W3(OH), H_2O masers trace a bipolar outflow (see Figure 6). There is a compact synchrotron source at the center of the flow, which suggests that shock excitation may be important in this region [*Reid et al.*, 1995]. Eight H_2O maser clusters have been monitored for proper motions and they all show outflows with expansion velocities of about 20 kms^{-1} (see Table 1).

3.2 Envelopes of Late–Type Stars

In late-type (red giant) stars, which exhibit substantial mass loss through winds, the masers of different species appear as nested shells where the conditions are optimum for each type of maser emission. The typical radii of the shells for SiO, H_2O and OH masers are 10^{14}, 10^{15} and $10^{16\text{-}17}$ cm, respectively. An example of the SiO distribution is shown in Figure 7. The SiO masers form in a shell only several stellar radii from the host star, near the boundary where silicon changes from the gas phase to the solid phase [see *Greenhill et al.*, 1995a]. Understanding this process is very important because the outflows from these stars replenish the interstellar medium in a cycle that leads to new star–formation.

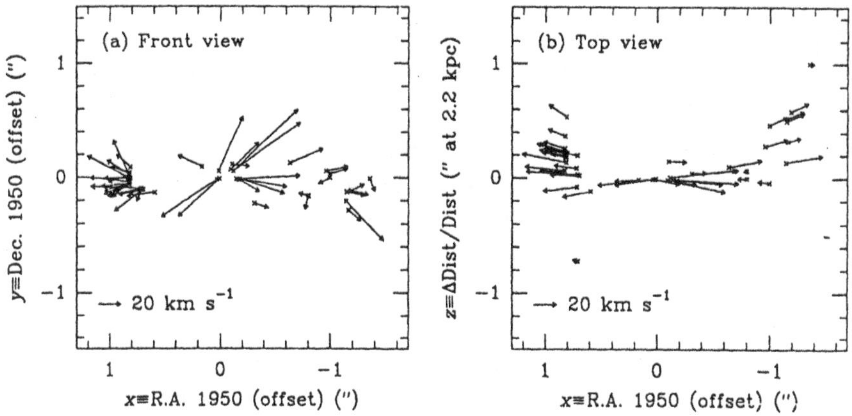

Figure 6: Map of the distribution of the H_2O masers near W3(OH). (a) Measured positions and proper motion based on a distance of 2.2 kpc. (b) Top view, with the assumption of constant outflow velocity. The motions clearly indicate a well collimated bipolar outflow [Alcolea et al., 1994].

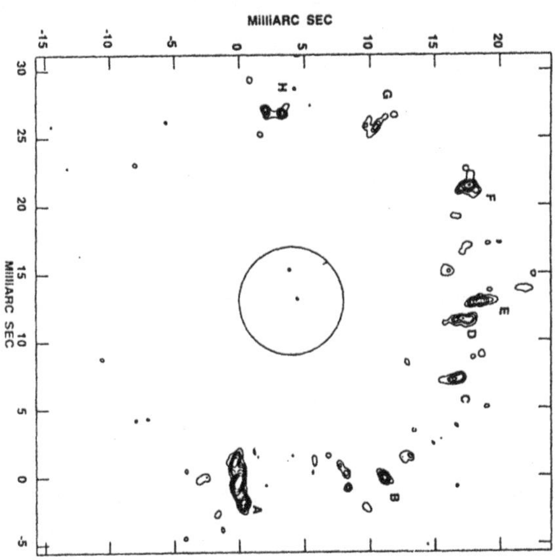

Figure 7: Map of the distribution of the SiO masers at 43 GHz towards the late–type star TX Cam. The circle in the center represents the stellar photosphere of diameter 3.6 AU [Diamond et al., 1994].

3.3 Supernova remnants

The association of OH masers in the 1720 MHz transition with supernovaremnants was postulated long ago [*Ball and Staelin*, 1968]. Recent VLA observations show an example of extensive maser emission in the envelope of a supernova remnant (see Figure 8). This type of maser may prove useful for tracing the extent of shock fronts, measuring gas motions and determining the magnetic field strength.

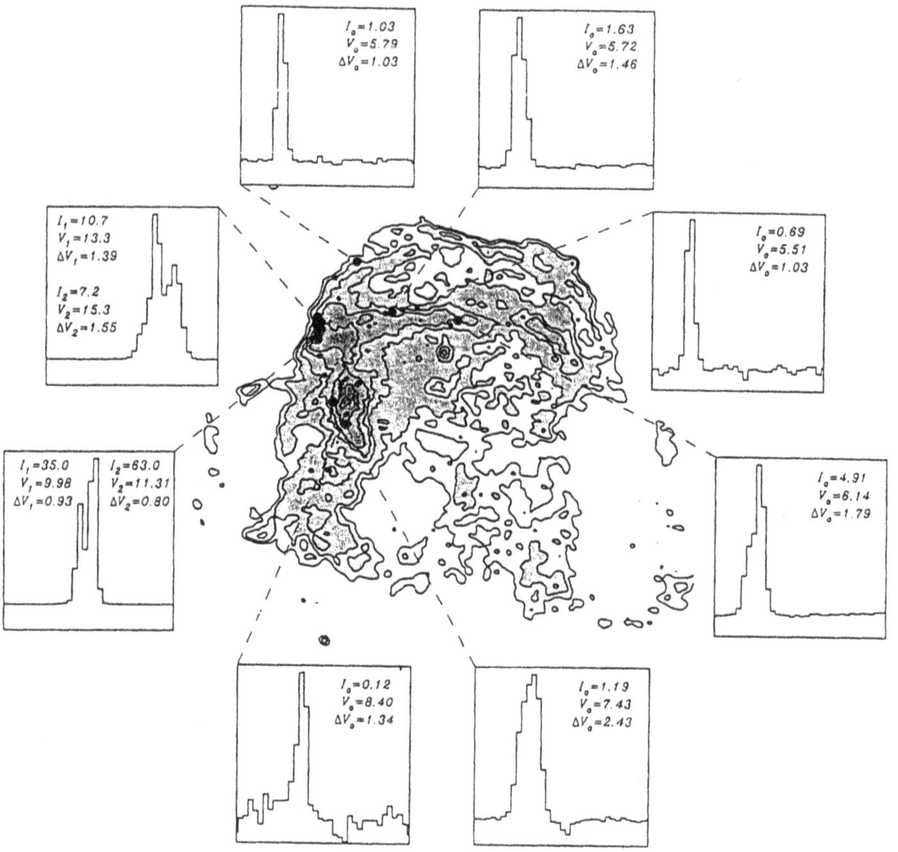

Figure 8: The distribution of masers superimposed upon the continuum image of the supernova remnant W28. Representative spectra are also shown [Frail et al., 1994a].

3.4 Active galactic nuclei

OH and H_2O masers have been discovered in the nuclei of many galaxies with active nuclei. OH masers have been found in galaxies with distances up to 100 Mpc and they may prove to be useful cosmological probes. About 5 percent of active galaxies show detectable H_2O maser emission. About 15 examples are currently known. The most interesting case of this phenomenon is in the galaxy NGC4258, a relatively nearby spiral galaxy that is a low luminosity type 2 Seyfert galaxy [*Urry* and *Padovani*, 1995]. The spectrum is shown in Figure 9. An unusual characteristic of the spectrum is the isolated groups of spectral features separated by ± 1000 kms^{-1} from the group of features near the systemic velocity of the galaxy (470 kms^{-1}). Observations with the VLBA [*Miyoshi et al.*, 1995] at high angular resolution (300 μas) and high spectral resolution (0.2 kms^{-1}) show that the emission comes from an elongated region (Figure 10). Furthermore, there is a monotonic progression of velocity with position among the maser spots along the major axis, as shown in Figure 11. This dependence is exactly the form expected for Keplerian motion of a disk orbiting a central mass, i.e., $V = \sqrt{GM/R}$, where G is the gravitational constant, M is the central mass, and R is the radius from the center of rotation of a maser with velocity V. The required binding mass is at least 3.5 x 10^7 $M_\odot$, and the inner part of the disk takes 800 yrs to complete one orbit. Furthermore, this mass must be contained within the inner edge of the disk, which has a radius of 0.13 pc. The corresponding mass density is at least 4 x 10^9 $M_\odot pc^{-3}$. This mass density is 10^9 times the stellar density in the vicinity of the Sun and 10^4 times the density of the densest known star cluster. The mostly likely origin of the central mass is a black hole [*Maoz*, 1995; *Kormendy* and *Richstone*, 1995].

Because of the precision of the VLBA measurements, the parameters of the disk can be measured with great accuracy. For example, the disk is unresolvable in the vertical direction and has a height of less than 10 μas. This may be the first direct detection of the outer part of a thin accretion disk, long thought to accompany a black hole. A thermally supported disk in hydrostatic equilibrium should obey the relation $H/R = c_s/V$ where H is the thickness of the disk and c_s is the sound speed. For the maser disk, $H/R < 0.0025$ so that the sound speed must be less than about 2.5 kms^{-1} and the temperature must be less than about 1000K. For such a disk the inward drift velocity is expected to be about 1 ms^{-1} [*Frank, King* and *Raine,* 1992] and the accretion rate is about 10^{-4} $M_\odot pc^{-3}$. This rate is sufficient to power the observed X–ray source (L_x = 4 x 10^{40} erg s^{-1}) [*Moran et al.*, 1995].

4 Magnetic fields

The line profile of a maser can be split by the Zeeman effect, which provides a mechanism for estimating the magnetic field strength. OH is a paramagnetic molecule and the splitting of the 1665 MHz transition is about 3.2 kHz/mG. Hence, a field of 1.7 mG is required for the splitting to equal the width of a typical line of 1 kms^{-1} or 5 kHz.

The Zeeman effect has proven to be an effective tool for measuring the magnetic fields in star-forming regions when interferometric measurements have been made to convincingly identify Zeeman pairs. Magnetic fields have been found to have strengths of about 5 mG [*Reid* and *Moran*, 1988]. Water is nonparamagnetic and the splitting of the 22 GHz line is only about 1.4 Hz/mG.

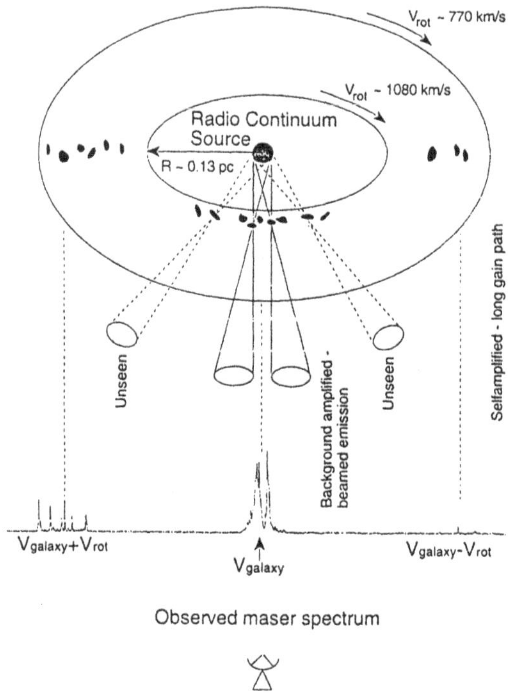

Figure 9: (bottom) Spectrum of the H_2O maser (1.35 cm wavelength) towards NGC4258. (top) Cartoon illustrating the geometry of the maser [Moran et al., 1995].

Hence, a magnetic field of 50 Gauss is required for a Zeeman slitting equal to the linewidth of typically 1 kms^{-1} or 74 kHz. However, the splitting can be detected by careful measurement of the difference in right and left circularly polarized spectra (see Figure 12). The field strength in H_2O masers are about 50 mG [*Fiebig* and *Güsten*, 1989]. For molecular clouds that collapse to form stars, flux freezing suggests that the magnetic field scales as the density squared. This gives a density ratio for gas in H_2O and OH masers of about 100, as expected on theoretical grounds for maser conditions. Efforts to measure magnetic fields in NGC4258 have only given limits on field strengths.

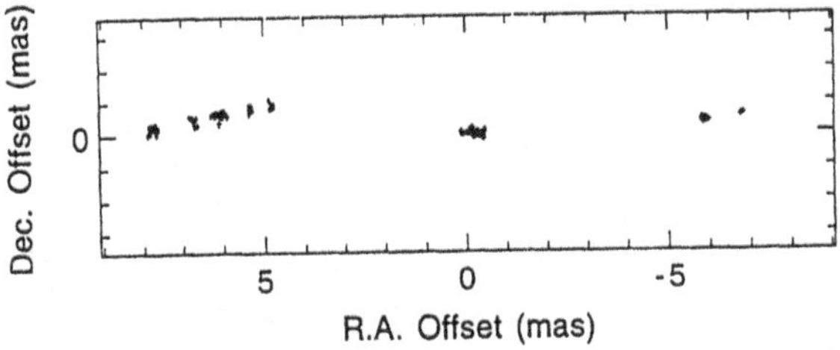

Figure 10: Distribution of maser towards NGC4258 which is attributed to emission from a disk viewed nearly edge–on. A slight warp is evident [Miyoshi et al., 1995].

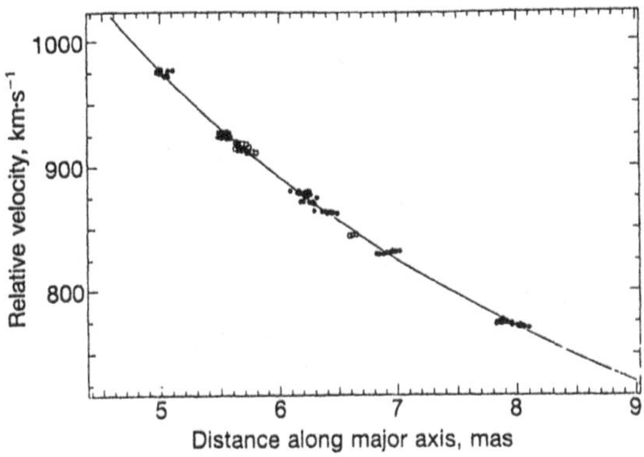

Figure 11: Distribution of line–of–sight velocities relative to the velocity of the galaxy [Miyoshi et al., 1995, Moran et al., 1995].

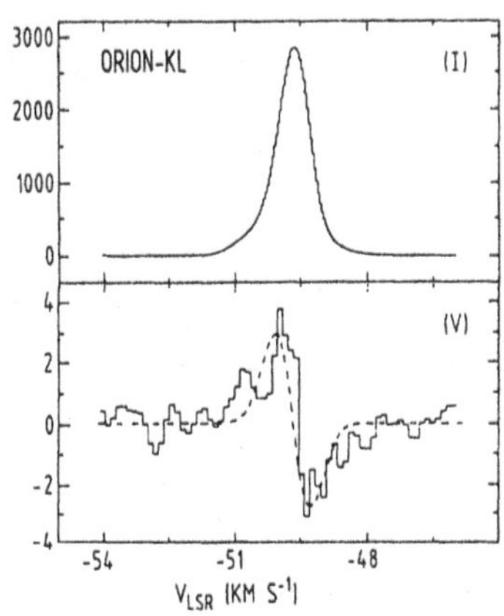

Figure 12. Stokes parameters I and V for the H_2O maser towards the star forming region Orion–KL. The "S" shaped V spectrum indicates the presence of a magnetic field of about 40 mG [Fiebig and Güsten, 1989].

5 Distance determinations

5.1 Expanding Envelopes around newly formed stars

The distance to a cluster of masers can be estimated by comparing the angular motions of the spots in the plane of the sky with their line-of-sight (Doppler) velocities. The method requires that the velocity field be isotropic. If the motions are random, then the distance can be estimated by the method of statistical parallax. If σ_μ is the dispersion in transverse angular velocities and σ_z is the dispersion in line–of–sight velocities, then the distance estimate that gives commensurable angular and linear velocities is $D = \sigma_z/\sigma_\mu$. The statistical error in the distance estimate is $\sigma_D/D \simeq 1/\sqrt{N}$ where N is the number of maser spots in the cluster.

Maser clusters associated with star formation regions are usually associated with expanding envelopes or winds (see Section 3.1). In this case, if the expansion velocity is independent of orientation, then positions, proper motions and velocities of the maser spots can be fit to an expanding source model. The distance is one model parameter, which assures commensurability between the angular and linear velocities.

Table 1: Distances of Galactic H_2O Masers Derived from Proper

Source	Fit (2)	N(3)	V_e(4) km s^{-1}	V(5) km s^{-1}	D_{PM}(6) kpc	D_K(7) kpc	χ(8)
Orion	M	26	18	8	0.5±0.1	0.7±1.0	-0.5
W51M	SP	27	-	63	0.7±1.5	5.0±2.0	0.8
W51N	SP	10	-	63	8.3±2.5	5.0±2.0	1.1
Sgr B2N	M	24	45	50	7.1±1.5	7.7±0.2	-0.4
Sgr B2M	M	27	35	26	6.5±1.5(9)	7.7±0.2	-0.9
W49N	M	105	18	7	11.4± 1.2	10.8±0.7	0.6
W3 (OH)	M	42	18	-48	- (10)	-	-
Cep A	SP	11	-	-11	0.5±0.2(9)	1.1±0.9	-1.4

(1)Primary references can be found in Moran [1993]. (2) SP = statistical parallax; M = model fit. (3) Number of maser features. (4) Outflow velocity in kms^{-1}. In Orion and W49N there are additional higher velocity flows. (5) Systemic velocity. (6) Proper motion distance. (7) Distance from kinematic model: v = 220 kms^{-1}, R_o = 7.8 kpc and random velocity = 4 kms^{-1}. (8) Weighted residual. (9) Preliminary estimate. (10) No distance estimate can be made because the outflow is well collimated and nearly in the plane of the sky (see Figure 6).

Distances to seven H_2O maser sources have been determined by these methods and the results are tabulated in Table 1. Because these masers originate from various places in the plane of our galaxy, the size of the galaxy can be estimated from a simple model of galactic rotation, i.e., constant circular motion at 220 kms^{-1}. The distance from the Sun to the Galactic Center, R_o, is estimated to be 7.8 ±0.6 kpc. See the review by *Reid* [1993] for a comprehensive discussion of the various measurements of R_o.

A preliminary estimate has been made of the distance to the maser cluster in the nearby galaxy M33 with the method of statistical parallax [*Greenhill et al.*, 1993]. These measurements are difficult because with a typical velocity of 30 kms^{-1} and an expected distance of 700 kpc, the angular motions areonly about 9 $\mu as\ yr^{-1}$ (see Eg 10).

5.2 Expanding Envelopes in Late–Type Stars

The OH masers at 1612 MHz arise preferentially from the parts of the spherical shell that are nearest to and farthest from the Earth (i.e., along the line of sight to the star). This happens because the OH is far enough from the star that the envelope has reached its terminal velocity and the velocity gradients along the line of sight are small, thus the amplification gain paths are large in these directions. The masers are pumped by infrared radiation from the star, which varies periodically. The flux densities of the maser features also vary periodically with the variations in stellar pump power. However, because of the light travel time, there will be a time lag, τ, between the "light curve" from the back side maser and the "light curve" from the front side maser. This lag is 8–80 days, corresponding to shell diameters of $10^{16\text{-}17}$ cm. If the size of the maser shell, θ (about 1 – 10"), is measured by interferometry, then the distance can be estimated by the simple relation, $D = c\tau/\theta$. Distances to about a dozen stars have been measured by this technique. The estimate of the size of the galaxy deduced from them gives $R_o = 8.8 \pm 0.9$ kpc (see summary in [*Moran*, 1993]).

5.3 Rotating disks in AGN

It should be possible to measure the proper motions of the maser in molecular disks in active galactic nuclei. Measurements are underway for NGC4258, where the proper motions are expected to be 32 x (6.4Mpc/D) $\mu as\ yr^{-1}$. However, a distance can also be estimated by comparing the centripetal accelerations of masers with the angular size and velocity of the disk. These accelerations manifest themselves as drifts in the line–of–sight velocities of spectral features with time as the systemic components on the near side of the disk move across the line of sight (see Figure 8) [*Greenhill et al.*, 1995b]. For NGC4258 this distance is 6.4±0.9 Mpc [*Miyoshi et al.*, 1995]. Optically determined distances range from 3.5 to 7 Mpc. Because the maser-motion technique is based solely on trigonometric calculations, it offers a method of estimating distance that is independent of more traditional methods, which are affected by absorption and other effects.

6 Interstellar scattering

The angular sizes of most masers are intrinsically very small. Their images can be "blurred" because of wavefront phase perturbations induced on the radiation as it passes through the turbulent ionized interstellar medium [*Rickett*, 1990]. The rms deviation in the index of refraction varies as λ^2 and hence the image size varies approximately as λ^2 ($\lambda^{11/5}$ for Kolmogorov turbulence) and also about as the square root of the secant of the galactic latitude and the square root of the distance. The scattering of maser emission has been studied extensively by *Gwinn et al.* [1989] and others. Scattering is particularly intense in the direction of the Galactic Center and amounts to about 1 " at 18 cm wavelength. Recently, *Frail et al.* [1994b] have found that the images of OH masers at 18 cm wavelength are anisotropic in the vicinity of the Galactic Center. This anisotropy may be due to the effect the magnetic field has on aligning the turbulence.

I thank my colleagues of CfA involved in spectral line VLBI measurements for simulating discussions (A. Argon, L. Greenhill, J. Herrnstein, K. Menten, M. Reid, and A. Trotter).

References

Alcolea, J., K. M. Menten, J. M. Moran, and M. J. Reid [1994], The Proper Motions of the H2O Masers Near W3(OH), in A. W. Clegg and G. E. Nedoluha, editors., *Astrophysical Masers, Lecture Notes in Physics, 412*, Springer–Verlag, Berlin, pages 225–228.

Ball, J. A., and D. H. Staelin [1968], Classification of OH Emission Sources, *Astrophys. J., (letters) 153*, L41–L46.

Barrett, A. H., and A. E. E. Rogers [1966], Observations of Circularly Polarized OH Emission and Narrow Spectral Features, *Nature, 210*, 188–190.

Bloemhof, E. E., M. J. Reid, and J. M. Moran [1992], Kinematics of W3(OH): First Proper Motions of OH Masers, *Astrophys. J., 397*, 500–519.

Clegg, A. W., and G. E. Nedoluha [1993], *Astrophysical Masers, Lecture Notes in Physics, 412*, Springer–Verlag, Berlin.

Cohen, R. J. [1989], Compact Maser Sources, *Rep. Prog. Phys, 52*, 881–943.

Deguchi, S., and W. D. Watson [1989], Interacting Masers and the Extreme Brightness of Astrophysical Water Masers, *Astrophys. J., (letters)*, 340, L17–L20.

Diamond, P. J., A. J. Kemball, W. Junor, A. Zensus, J. Benson, and V. Dhawan [1994], Observation of a Ring Structure in SiO Maser Emission from Late–Type Stars, *Astrophys. J., 430*, L61–L64.

Evans, N. J., R. E. Hills, O. E. H. Rydbeck, and E. Kollberg [1972], Statistics of the Radiation from Astronomical Masers, *Phys. Rev. A, 6,* 1643–1647.

Elitzur, M. [1992a], *Astronomical Masers,* Kluwer Academic Publishers, Dordrecht.

Elitzur, M [1992b], Astronomical Masers, *Annu. Rev. Astron. Astrophys, 30,* 75–112.

Elitzur, M., D. J. Hollenbach, and C. F. McKee [1989], H2O Masers in Star Forming Regions, *Astrophys. J., 346,* 983–990.

Fiebig, D., and R. Güsten [1989], Strong Magnetic Fields in Interstellar H2O Maser Clumps, *Astr. Astrophys., 214,* 333–338.

Frail, D. A., W. M. Goss, V. I. Slysh [1994a], Shock–Excited Maser Emission from the Supernova Remnant W28, *Astrophys. J.,l 424,* L111–L113.

Frail, D. A., P. J. Diamond, J. M. Cordes, and H. J. Van Langevelde [1994b], Anisotropic Scattering of OH/IR Stars Toward the Galactic Center, *Astrophys. J., 427,* L43–L46.

Frank, J., A. King, and D. Raine [1992], *Accretion Power in Astrophysics,* Cambridge Univ. Press, Cambridge.

Goldreich, P., and J. Kwan [1974], Astrophysical Masers. V. Pump Mechanism for H2O Masers, *Astrophys. J., 191,* 93–100.

Greenhill, L. J., J. M. Moran, M. J. Reid, K. M. Menten, and H. Hirabayashi [1993], Microarcsecond Proper Motions of Extragalactic Water Vapor Masers in M33, *Astrophys. J., 406,* 482–488.

Greenhill, L. J., F. Colomer, J. M. Moran, D. C. Backer, W. C. Danchi, and M. Bester [1995a], Interferometric Observations of the SiO Masers and Dust Shell of VXC Sagittarii, *Astrophys. J., 449,* 365–375.

Greenhill, L. J., C. Henkel, R. Becker, T. L. Wilson, and J. G. A. Wouterloot [1995b] Centripetal Acceleration within the Subparsec Nuclear Maser Disk in NGC4258, *Astr. Astrophys., 304,* 21–33.

Gundermann, E. [1965] Observations of the Interstellar Hydroxl Radical, Ph.D. thesis, Harvard University, page 50.

Gwinn, C. R., J. M. Moran, M. J. Reid, and M. H. Schneps [1988], Limits on Refractive Interstellar Scattering Towards Sgr B2, *Astrophys. J., 330,* 817–827.

Kormendy, J. and D. Richstone [1995], Inward Bound–The Search for Supermassive Black Holes in Galactic Nuclei, *Annu. Rev. Astron. Astrophys., 33,* 581–624.

Maoz, E. [1995], A Stringent Constraint on Alternatives to a Massive Black Hole at the Center of NGC4258, *Astrophys. J., (letters), 447,* L91–L94.

Menten, K. [1994], Dense Molecular Gas in Star–Forming Regions–The Importance of

Submillimeter Observations, in T.L. Wilson and K.J. Johnston, editors, *The Structure and Content of Molecular Clouds*, Lecture Notes in Physics, 439, Springer–Verlag, Berlin, pages 109–122.

Miyoshi, M., J. Moran, J. Herrnstein, L. Greenhill, N. Nakai, P. J. Diamond, M. Inoue, [1995], Evidence for a Massive Black Hole from High Rotational Velocities in a Sub-parsec Region of NGC4258, *Nature*, 373, 127-129.

Moran, J. M., L. G. Greenhill, J. R. Herrnstein, P. J. Diamond, M. Miyoshi, N. Nakai, and M. Inoue [1995], Probing Active Galactic Nuclei with H2O Megamasers, *Proc. Nat. Acad. Sci., 92*, 11427–11423.

Napier, P. J., D. S. Bagri, B. G. Clark, A. E. E. Rogers, J. D. Romney, A. R. Thompson, and R. C. Walker [1994], The Very Long Baseline Array, *Proc. I.E.E.E., 82*, 658–672.

Reid, M. J. [1993], The Distance to the Center of the Galaxy, *Ann. Rev. Astro. and Astrophys., 31*, 345–372.

Reid, M. J., Argon, A. L., C. R. Masson, K. M. Menten, and J. M. Moran [1995], Synchrotron Emission from the H2O Maser Source in W3(OH), *Astrophys. J.*, 443, 238–244.

Reid, M. J. and J. M. Moran [1988], Astronomical Masers, in G. L. Verschuur and K. I. Kellermann, editors, *Galactic and Extragalactic Radio Astronomy*, Springer–Verlag, Berlin, pages 255–294.

Rickett, B. J. [1990], Radio Propagation Through the Turbulent Interstellar Plasma, *Ann. Rev. Astron. Astrophys., 28*, 561–605.

Shu, F. H., F. C. Adams, and S. Lizano [1987], Star Formation in Molecular Clouds: Observation and Theory, *Annu. Rev. Astron. Astrophys., 25*, 23–81.

Urry, C. M., and P. Padovani [1995], Unified Schemes for Radio–Loud Active Galactic Nuclei, *Pub. Ast. Soc. Pac., 107*, 803–845.

Watson, W. D. [1994], Interpretations of Observations of Astrophysical Masers, in T.L. Wilson and K.J. Johnston, editors, *The Structure and Content of Molecular Clouds, Lecture Notes in Physics, 439*, Springer–Verlag, Berlin, pages 109–122.

Weaver, H., D. R. W. Williams, N. H. Dieter, and W. T. Lum [1965], Observations of a Strong Unidentified Microwave Line and Emission from the OH Molecule, *Nature*, 208, 29–31.

Wilson, T. L., and K. J. Johnston [1994], *The Structure and Content of Molecular Clouds, Lecture Notes in Physics 439*, Springer–Verlag, Berlin.

PERSONAL COMMUNICATION SERVICES - TECHNOLOGY AND HEALTH CONCERNS: IS THERE A COMMON SOLUTION?

Maria A. Stuchly

1 Introduction

With the expansion of current use and anticipated further increases in the use of cellular telephones and other personal communication services (PCS), there have been an interest and considerable research effort devoted to interactions between antennas on hand-sets and the human body. The motivation for research in this area is due to two factors. From the technology point of view, as many of the PCS devices are used close to the human body, the latter results in a deterioration of the antenna performance and may therefore either require operation with increased power or other means that prevent the deterioration of communication. From the health point of view, concerns are expressed that the deposition of microwave power in some locations may be excessive and therefore lead to detrimental health effects.

Considerable progress has been made in understanding the interactions of RF and microwave fields with living systems. This understanding and agreement among the majority of scientists and regulators has resulted in very similar or nearly the same recommendations regarding safe exposure levels in many national standards and international guidelines [*WHO*, 1993]. Wireless commun- ications present new challenges in terms of health effects. There are two main issues. The first is microwave power from a cellular telephone. In some cases up to 50% of the device output power may be deposited in the user's body [*Stuchly*, 1994] when the antenna is placed very close to the head. The power is localized within a small volume. The second issue relates to new technologies which improve the efficiency and channel utilization through digital modulations (e.g. time division multiple access - TDMA, or code division multiple access - CDMA), and use of frequency shift keying (FSK) or phase shift keying (PSK). This results in the amplitude (pulse) modulation at extremely low frequencies (ELF) of microwave signal. There is very little biological data on effects of such fields.

In recent years there has been considerable effort devoted to modeling interactions of cellular telephones with their user body, particularly the head and the hand. Most of the effort has been motivated by the health concerns, but also new designs of antennas have been proposed and analyzed. There is little doubt that the goals of technology and health concerns are leading to the same solutions, namely design of hand-sets, and particularly antennas, that deposit less microwave power in the user's body. This paper reviews modeling of interactions of RF/microwave fields from cellular telephones with the user's head and hand from the perspective of the mentioned commonality of goals. For completeness, brief reviews are given of biological effects and of current developments in the finite difference time domain method (FDTD), a method of choice for analyses of the problems, that are relevant to the issue addressed.

2 Biological effects of microwaves

Biological effects of radio and microwave fields have been extensively investigated as documented in several reviews [*WHO*, 1993; *Saunders, et. al.*, 1991; *Bernhardt*, 1992; *Stuchly*, 1987 *and ANSI C95.1*-1992]. With the help of dosimetric modeling, thresholds of the average specific absorption rate (SAR) or rates of energy deposition in tissue in watts per kilogram have been established for various effect in animals. Some examples follow, however a comprehensive review with a full statistical analysis is not attempted here. Data quoted is from the published reviews [*WHO*, 1993; *Mumford, et. al.*, 1994 *and Bernhardt*, 1992]. Radio frequency radiation is teratogenic in rodents at SARs that approach lethal levels, and a threshold for induction of birth defects is associated with the maternal core temperature of 41 - 42^{0}C [*WHO*, 1993; *Saunders, et. al.*, 1991 *and Stuchly*, 1987]. Chronic exposure of rats during gestation at 2.5 W/kg was reported to result in lowered fetal body weight at weaning [*WHO*, 1993]. Temporary sterility in male rats occurred at an SAR of 5.6 W/kg, which produced a core temperature of 41^{0}C [*WHO*, 1993]. Decreases in operant and learned-behavioral responses occur at SAR = 2.5 W/kg in the rat and at 5.0 W/kg in the rhesus monkey. Some types of behavior are affected at SARs approximately 25-50% of the resting metabolic rate. These behavioral changes are reversible with time, after exposure ceases. Changes in the endocrine system and blood chemistry at SARs greater than 1 W/kg and in hematologic and immunologic systems at SARs equal and greater than 0.5 W/kg, for prolonged exposures, appear to be associated with thermal stress [*WHO*, 1993 *and Stuchly*, 1987]. One group of researchers reported that chronic exposure at SARs of 2-3 W/kg resulted in cancer promotion or co-carcinogenesis in mice [*Szmigielski, et. al.*, 1982]. The effect was similar to that caused by chronic stress. Neurons in the central nervous system were altered by chronic exposure at 2 W/kg [*WHO*, 1993]. All these effects appear to be associated with thermal load due to RF exposure and are characterized by thresholds [*WHO*, 1993; *Saunders, et. al.*, 1991; *Bernhardt*, 1992; *Stuchly*, 1987 *and ANSI C95.1*-1992].

For whole-body exposures to RF/MW radiation, the experimental data currently available strongly suggests that biological effects in mammals occur at average SARs of about 1 to 4 W/kg. The database is well established and consistent. This fact has been reflected in more recent exposure standards such as the US ANSI C95.1-1992 [*ANSI C95.1*-1992] Canadian Safety Code, 1991 Revision, [*Safety Code* 6, 1991] the UK, NRPB Standard [*NRPB*, 1989]. Each of these documents contain a reference either in the text or in the accompanying publication to the same database [*WHO*, 1993; *Saunders, et. al.*, 1991; *Bernhardt*, 1992 *and Stuchly*, 1987]. The main limitation of the data base is a lack of well-performed chronic studies, and the paucity of data for ELF amplitude modulated fields. It needs to be emphasized that in the case of cellular telephones and other PCS devices, exposure is localized, and whole-body average SARs are not directly applicable.

Biological effects have been observed at RF and MW fields amplitude modulated at ELF at SAR levels below thresholds for effects for continuous waves [*WHO*, 1993; *Bernhardt*, 1992 *and Stuchly*, 1990]. Many of these effects are the same or similar to effects observed for ELF electric and magnetic fields. The observed effects are usually field frequency and intensity specific. They tend to occur within relatively narrow ranges of both field parameters, and are dependent on other physical and physiological characteristics of the exposed biological system. Many of these parameters have not been fully identified and characterized. The interaction mechanisms remain unknown. The scientific database is relatively limited in this area. There is little indication that any of these effects is detrimental to human health, but the potential importance of these effects should be critically evaluated.

3 Models used in analyses

Several models of the human head have been analyzed. They range from simple models of the human head such as a homogeneous sphere [*Toftgard et. al.*, 1993] or a homogeneous head shaped volume [*Chuang*, 1995], to heterogeneous, anatomically correct models based on the magnetic resonance imaging (MRI) [*Jensen and Rahmat-Samii*, 1995; *Dimbylow and Mann*, 1994; *Gandhi*, 1995; *Martens et. al.*, 1995; *Okoniewski and Stuchly*, 1996] and the computer tomography (CT), [*Okoniewski and Stuchly*, 1996].

Relatively few studies modeled the hand holding a handset [*Jensen and Rahmat-Samii*, 1995; *Gandhi*, 1995; *Douglas et. al.*, 1995; *Okoniewski and Stuchly*, 1996]. In all cases the hand was modeled as a simple two-layer structure (box-like) consisting of muscle and bone.

The so called "canonical" models of the human head for evaluation of SARs have been developed in Europe as a part of the COST 244 WG3 project. Two main head geometries are a box and a sphere (both either homogeneous or layered).

Several studies dealt with a "generic" model of a cellular telephone at approximately 900 MHz consisting of a monopole on a metal box 15 cm long with a cross-section of 5 to 6 cm by 1.3 to 2.5 cm [*Jensen and Rahmat-Samii*, 1995; *Toftgard et. al.*, 1993; *Dimbylow and Mann*, 1994; *Martens et, al.*, 1995; *Okoniewski and Stuchly*, 1996]. A few commercially available telephones were analyzed, but details of their antenna type and box geometry were not given [*Gandhi*, 1995]. A few antenna designs whose performance is less affected by the human head and which produce lower SAR in the head were also numerically modeled [*Jensen and Rahmat-Samii*, 1995; *Douglas et. al.*, 1995; *Okoniewski and Stuchly*, 1996].

4 Method of analysis

Although other methods of analysis have been used, the FDTD method has become a method of choice in modeling interactions of cellular telephones with the user's body. This is due to both the inherent features of the method and the rapid progress in its development in the last few years. The features that are particularly attractive is the method simplicity and computational efficiency, that facilitate obtaining results within the whole frequency range of interest in one computation and simultaneous evaluation of the antenna performance and SARs. Furthermore, relatively large computational spaces can be modeled with reasonably fine resolution on a typical workstation, while really large problems with fine resolution can be analyzed on supercomputers including those with parallel processors. The method was proposed by *Yee* [1966] and its detailed description as well as current progress can be found in a book by *Taflove* [1995].

One of the special recently developed features is a perfectly matched layer (PML) proposed by *Berenger* [1994] and used as the absorbing boundary to ensure reflections below at least 40 dB from the termination of the computational space. Other absorbing boundary conditions can and have been used, but a properly chosen PML offers superior performance at the lowest expense of the computational expenses.

In the cases where the metal surfaces did not coincide with the mesh, which is frequently the case when the actual geometry of the antenna and handset is considered, a special algorithm can be used that would account for the actual exact position of these surfaces [*Anderson et. al.*, 1995]. To model fields in selected subregions such as the antenna feed or a part of the head a subgridding algorithm can be applied [*Okoniewski et. al.*, 1995].

The far-field antenna pattern is obtained from the computed near-field electric and magnetic field vectors on a box enclosing the modeled structures using the field equivalence principle [*Taflove*, 1995]. The antenna efficiency is computed as:

$$\eta = \frac{P_{rad}}{P_{rad} + P_{abs}} \tag{1}$$

where P_{rad} is calculated by integrating the normal component of the Poynting vector over the surface of the box used for calculations of the near-far-field transformation. The power absorbed (in the head model and hand) is:

$$p_{abs} = \frac{1}{2}\int_{v} \sigma |E_t|^2 dv \tag{2}$$

where E_t is the total electric field in a voxel of tissue. The SAR is computed as:

$$SAR = \frac{\sigma}{2\rho}|E_t|^2 \tag{3}$$

where: σ is the tissue conductivity, and ρ the tissue specific density.

5 Results and discussion

Modeling results obtained in our laboratory are discussed in more detail in this review, as they are quite representative of the research in this area. They are complimented by results obtained by others for comparison purposes or where the data address a different problem or differ from ours.

(a)

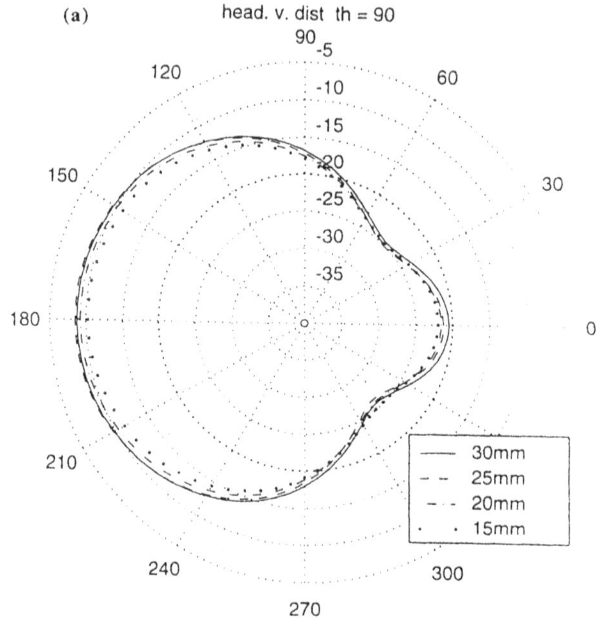
head. v. dist th = 90
90
-5
-10
-15
-20
-25
-30
-35
120
60
150
30
180
0
210
240
270
300
30mm
25mm
20mm
15mm

(b)

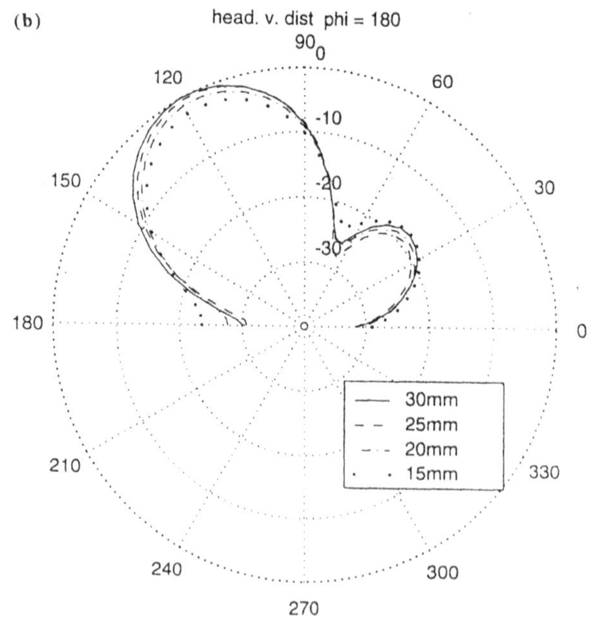
head. v. dist phi = 180
90
0
-10
-20
-30
120
60
150
30
180
0
210
330
240
270
300
30mm
25mm
20mm
15mm

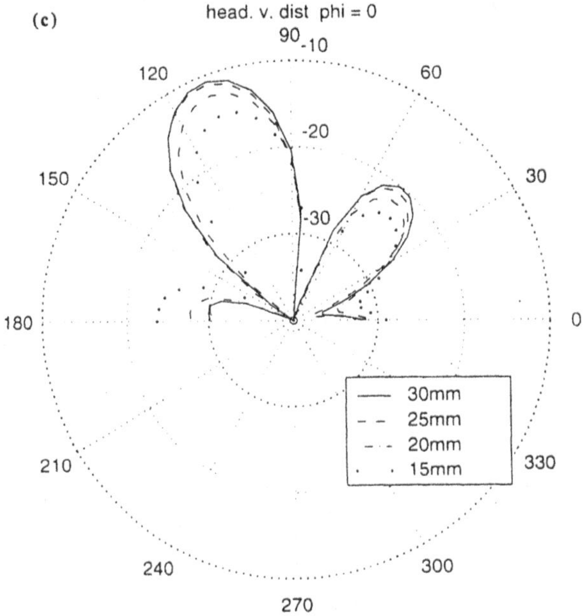

Fig. 1 Radiation pattern for a monopole on a dielectric covered metal box at 915 MHz - for various distances between the antenna and the head [Okoniewski and Stuchly, 1996], (a) ϕ plane with $\theta = 90°$, (b) θ plane with $\phi = 180°$ (away from the head), and (c) θ plane with $\phi = 0°$ (towards the head).

Antenna Characteristics

The main antenna characteristics that are affected by the human body are: the radiation pattern, input impedance and efficiency. The radiation patterns of typical antennas currently used in cellular telephones, such as various monopoles, usually a quarter-wavelength monopole, are significantly changed in the presence of the head in their vicinity. Figure 1 shows the radiation patterns in three planes for a monopole at 915 MHz on a metal handset box of 13.5 cm (length) x 5 cm (width) x 2 cm (depth). The box is covered with a dielectric insulator 2 mm thick with $\varepsilon_r' = 2.5$. An accurate model of the head is based on a MRI scan and has a resolution of 3.4 x 3.4 x 4.7 mm [*Zubal et. al.*, 1994]. Table 1 summarizes the tissues segmented and the dielectric constant and conductivity allocated.

Tissue	ε '	σ (S/m)
Skin	35	0.6
Skull	8	0.11
Spinal cord	49	1.1
Spine	8	0.11
Brain - white matter	38	0.8
Brain - grey matter	49	1.1
Jaw bone	8	0.11
Muscle	58	1.4
Parotid gland	55	1.0
Lacrimal glands	55	1.0
Spinal canal	72	2.1
Tongue	55	1.0
Pharynx	35	0.6
Esophagus	35	0.6
Nasal septum	35	0.6
Fat	6	0.08
Blood	62	1.5
CSF	78	2.1
Eye - sclera	66	1.7
Eye - humour	74	2.0
Lens	44	0.8
Bone marrow	42	0.8
Cartilage	35	0.6
Pituitary gland	55	1.0
Ear bones	35	0.6
Trachea	35	0.6

Table 1 Dielectric properties of the tissues used in the head model [*Okoniewski & Stuchly*, 1996].

Results similar to these were also reported by *Jensen* and *Rahmat-Samii* [1995]. Another important issue that was addressed is what head models were adequate for antenna pattern evaluation [*Jensen* and *Rahmat-Samii*, 1995; *Okoniewski* and *Stuchly*, 1996]. Figure 2 illustrates that a box model of the head affected the antenna pattern quantitatively quite differently from the realistic interaction shown in Fig. 2 [*Okoniewski* and *Stuchly*, 1996]. On the other hand, a spherical model of the head gave reasonable approximation.

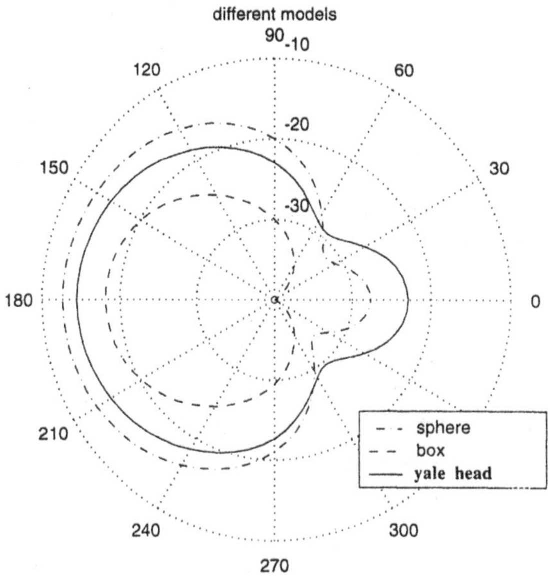

Fig. 2 Radiation patterns in (a) ϕ plane with $\theta \simeq 90°$, and (b) θ plane with $\phi = 180°$ for a monopole at 915 MHz on a dielectric covered metal box for a three layered tissue box for a sphere and an anatomically correct head model [Okoniewski and Stuchly, 1996].

The input impedance of the antenna is affected by the human body in its vicinity, but the effect is relatively small [*Jensen and Rahmat-Samii*, 1995]. Similarly, with the antenna pattern a spherical model of the head provides adequate simulation of the human head.

Additionally to influence of the user's head on the performance of a cellular telephone, the hand holding the handset perturbs all the characteristics. Perhaps the most dramatic effect is on the antenna efficiency, as illustrated in Table 2.

Model	**η (%)**	**P_{abs} (W)**	**SAR (W/kg)**			
			Head Peak	**Hand Peak**	**1 g head**	**10 g head**
Head & Hand	65	0.35	-	25.3	1.5	0.8
Head, no hand	61	0.39	2.9	-	2.1	1.1

Table 2 Effect of the head and hand on the antenna characteristics and SAR

SAR in the Head

Numerous evaluations of the absorbed power and its distribution (SAR) are reported by various investigators. Illustrative results are summarized in Table 3 for cellular telephones operating in the 900 MHz band. Similar, but fewer, results are also reported in the 2 GHz band [*Dimbylow and Mann,* 1994]. Several general observations can be made. They apply to hand sets with monopole or similar linear antennas. The total absorbed power and maximum SARs depend critically on the distance between the antenna and handset. The quality of the head model plays also a critical role, particularly the inclusion of the ear [*Okoniewski and Stuchly,* 1996]. This feature of the model, together with distance, likely explains lower SARs reported by some investigators [*e.g. Gandhi,* 1995] (consult Table 3). Even the spherical model of the head with 3 tissue layers does not provide satisfactory results with respect to SARs. However, the spherical model as well as a more realistic model but without the ear, both give higher values of SAR. This may be useful for initial investigations to determine either the "exaggerated worst case" SAR, or to develop antennas with lower SARs (and therefore also an improved communications performance).

Reference	Antenna - Head Distance	P_{abs} (W)	SAR peak (W/kg)	SAR 1 g (W/kg)	SAR 10 g (W/kg)
Dimbylow & Mann [1994]	1.4 cm	0.385	-	3.62	3.04
	2.0 cm	0.315	-	2.14	1.67
Jensen&Rahmat-Samii [1995]	1.97 cm	0.535	2.06	-	-
Gandhi, [1995] *	-	0.1 - 0.28	1.6 - 6.0	0.26 - 1.2	-
Okoniewski & Stuchly, [1996]	1.5 cm	0.40	3.5	2.6	1.8
	2.0 cm	0.33	2.4	1.7	1.4
	3.0 cm	0.27	1.9	1.2	0.9

* 10 commercially available telephones

Table 3 Representative SARs computed by various investigators for a monopole on a metal box in the 900 MHz band, power to the antenna 1W. (Data are not time averaged.)

The vertical position of the antenna (monopole) results in higher SARs than any other realistic position when the user is also listening [*Jensen and Rahmat-Samii,* 1995]. The high values of SAR in the eye [*Dimbylow and Mann,* 1994] occur only with rather unrealistic telephone placements at least by an adult but perhaps more likely for a child at play. In this case, the actual power transmitted in the stand-by mode of operation needs to be considered.

New Antenna Designs

Various patch antennas were investigated by ***Jensen and Rahmat-Samii*** [1995] and a diversity antenna was investigated by ***Douglas et. al.***, [1995]. Some of the antennas were shown to be less affected by the presence of the head as summarized in Table 4, and illustrated in Fig. 3 showing the radiation pattern. These improved designs while depositing less power in the head and producing lower local SARs are still affected by the user's hand. Recently, a well known principle of simulation of an open circuit in the ground plane was used to disconnect the part of the handset that is grasped by the hand from that part that serves as the ground plane for the antenna [***Okoniewski and Stuchly***, 1996]. The improvement of the effect on the antenna pattern is shown in Fig. 4, while Table 4 shows other features of this improved design.

Antenna Type	**Antenna Head Distance**	η (%)	**Peak SAR (W/kg)**
Side mounted PIFA	1.97 cm	0.34	2.07
Back mounted PIFA	1.97 cm	0.52	0.90

Table 4 Power absorbed and SARs for patch antennas [Jensen and Rahmat-Samii, 1995]

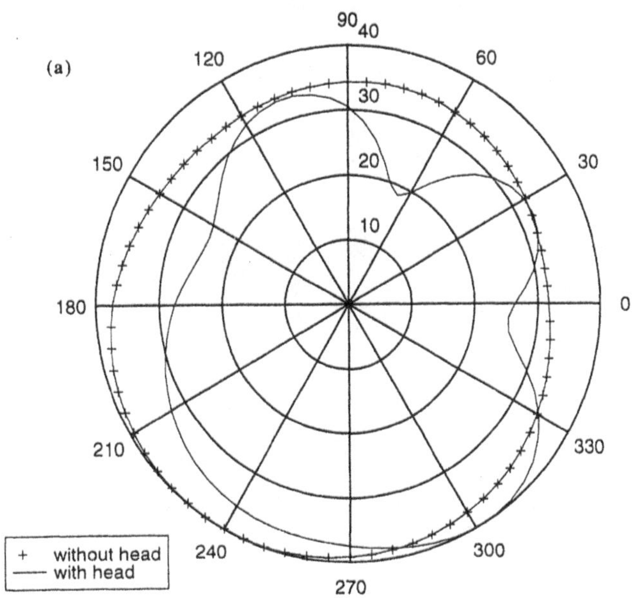

Fig. 3a Radiation pattern for a polarization diversity patch antenna [Douglas, et. al., 1995], (a) vertically-polarized mode.

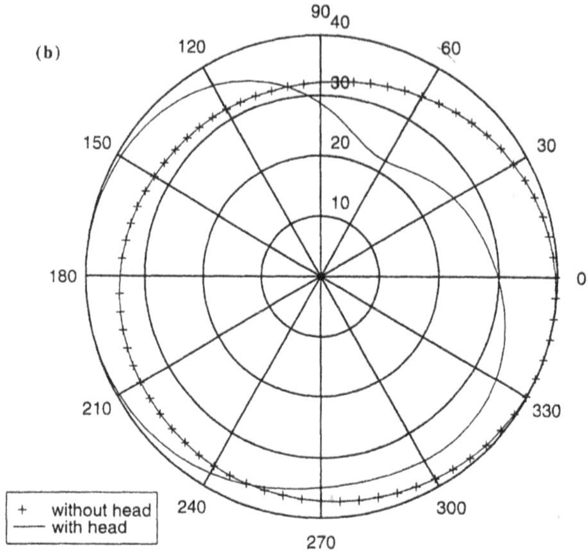

Fig. 3b Radiation pattern for a polarization diversity patch antenna [Douglas, et. al., 1995], (b) horizontally polarized mode.

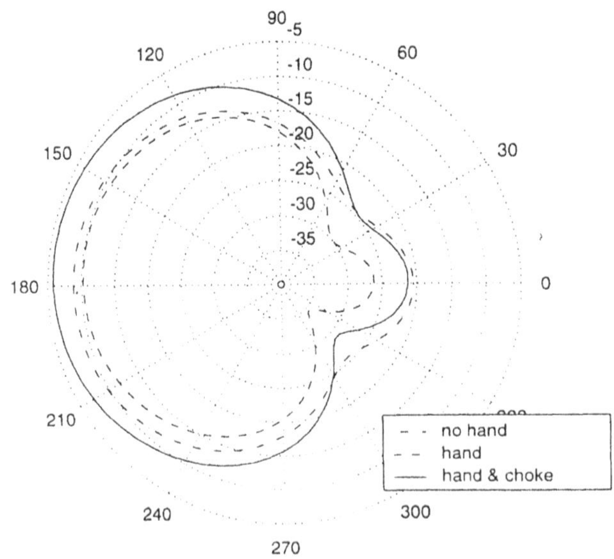

Fig. 4 Radiation pattern for a monopole on a dielectric covered metal box of 915 MHz 2 cm away from the head model, without the hand, with the hand, and with the hand and the new current interruption design.

6 Conclusions

In the last few years, with the growth of the use of PCS devices, particularly cellular telephones, considerable research effort has been devoted to evaluation of the influence of the telephone user's body on the communication performance and possible health effects due to the energy deposition in the head. The considerations of the health effects that have received a lot of attention and have been outlined in this paper are limited to the SARs as compared to those specified in health-safety recommendations. The other consideration of health effects relates to the type of modulation, that produces amplitude modulation at ELF used in the new generation of devices. No papers have been published in this area recently and the existing data base is very limited.

For low power cellular telephones, as those in use today, the SARs are very low compared to the standards and the existing evidence does not indicate any detrimental health effects. However, considerable amounts of output power, up to 50% in some cases, are deposited in the user's body. This leads to commonality of goals from the communications and health points of view, namely limiting the power deposited in the body. Improvements can be achieved through new designs of the antennas and of the handset.

7 Acknowledgments

This work has been supported by the Natural Sciences and Engineering Research Council of Canada, BC Hydro, TransAlta Utilities, industrial research chair.

References

ANSI C95.1 [1992], American National Standard Safety Levels with respect to human exposure to radio frequency electromagnetic fields, 300 kHz to 100 GHz, *The Institute of Electrical and Electronics Engineers, Inc.*, New York, N.Y.

Anderson, J., M. Okoniewski and S. S. Stuchly, [1995] "Practical 3-D contour/staircase treatment of metals in FDTD", *IEEE Microwave and Guided Wave Letters.*

Berenger, J-P., [1994], "A perfectly matched layer for the absorption of electromagnetic waves", *J. Comp. Phys.*, vol. 114, pp. 185-200.

Bernhardt, J. H., [1992], "Non-ionizing radiation safety: radiofrequency radiation, electric and magnetic fields", *Phys. Med. Biol.*, vol. 37, pp. 807-844.

Chuang, H. R., [1995], "Human operator coupling effects on radiation characteristics of a portable communication dipole antenna", *IEEE Trans. Antennas and Prop.*, vol. 42, pp. 556-560.

Dimbylow, P. J. and S. M. Mann, [1994], "SAR calculations in an anatomically realistic model of the head for mobile communication transceivers at 900 Mhz and 1.8 GHz", *Phys. Med. Biol.*, vol. 39, pp. 1537-1553.

Douglas, M., M. Okoniewski and M. A. Stuchly, [1995], "Modeling of wireless personal communication transmitters in the presence of the user's body", *Wireless '95 - 7th. Int. Conf. on Wireless Communication*, Calgary, Alberta, pp. 77-83.

Gandhi, O. P., [1995], "Some numerical methods for dosimetry: Extremely low frequencies to microwave frequencies", *Radio Science*, vol. 30, pp. 161-177.

Jensen, M. A. and Y. Rahmat-Samii, [1995], "EM interaction of handset antennas and a human in personal communications", *Proc. IEEE*, vol. 83, pp. 7-17.

Martens, L., J. De Moerloose and D. DeZutter, [1995], "Calculation of the electromagnetic fields induced in the head of an operator of a cordless telephone", *Radio Science*, vol. 30, pp. 283-290.

Mumford, R., Q. Bolzano and T. Taga, [1994], "Land mobile systems II: Pagers, portable phones and safety", *Mobile Antenna Systems Handbook*, Editors: K. Fujimoto and J. R. James, Artech House, Inc., pp. 175-273.

NRPB, [1989], "Guidance as to restriction on exposure to time varying electromagnetic fields and the 1988 recommendations of the International Non-ionizing Committee", *National Radiological Pro Board*, Chilton, Didcot, NRPB-GS-11, (London, HMSO).

Okoniewski, M., E. Okoniewska and M. A. Stuchly, [1995], "3D Subgridding algorithm for FDTD", *Digest of IEEE Antennas and Prop. Symp.*, vol. 1, pp. 232-235.

Okoniewski, M. and M. A. Stuchly, [1996], "A study of the handset antenna and human body interaction" *IEEE Trans. Microwave Theory and Tech.*, vol. 44, (accepted).

Safety Code 6, [1991], "Limits of exposure to radiofrequency fields at frequencies from 10 kHz - 300 GHz", *National Health and Welfare (Canada)* EHD-TR-160, Cat. # H46-2/90-160E.

Saunders, R. D., C. I. Kowalczuk and Z. J. Sienkiewicz, [1991], "The biological effects of exposure to non-ionizing electromagnetic fields and radiation: III Radiofrequency and microwave radiation", *National Radiological Protection Board*, Chilton, Didcot, NRPB-R241.

Stuchly, M. A., [1987], "Proposed revision of the Canadian recommendations on radiofrequency-exposure protection", *Health Phys.*, vol. 53, pp. 649-665.

Stuchly, M. A., [1990], "Applications of time-varying magnetic fields in medicine", *Crit. Rev. Biomedical Engineering*, vol. 28, pp. 89-124.

Stuchly, M. A., [1994], "Wireless communications and the safety of the user", *Int. J. Wireless Information*, vol. 1, pp. 223-228.

Szmigielski, S., A. Szudzinski, A. Pierraszek, B. Bielec, M. Janiak and J. K. Wrembel, [1982], "Accelerated development of spontaneous and benzopyrene-induced skin cancer in mice exposed to 2450 MHz microwave radiation", *Bioelectromagnetics*, vol. 3, pp. 179-191.

Taflove, A., [1995], "Computational electrodynamics: The finite-difference time-domain method", *Artech House Publications.*

Toftgard, J., S. Hornsleth and J. Bach Anderson, [1993], "Effects on portable antennas of the presence of a person", *IEEE Trans. Antennas and Prop.*, vol. 41, pp. 739-746.

WHO, [1993], "Electromagnetic Fields (300 Hz - 300 GHz)", *Environmental Health Criteria* 137, World Health Organization, Geneva.

Yee, K. S., [1966], "Numerical solution of initial boundary value problems involving Maxwell's equations in isotropic media", *IEEE Trans. Antennas and Prop.*, vol. 14, pp. 302-307.

Zubal, I. G., C. R. Harrell, E. O. Smith, Z. Rattner, G. R. Gindi, and P. H. Hoffer, [1994], "Computerized three-dimensional segmented human anatomy", *Med. Phys. Biol.*, vol. 21, pp. 299-302.

The manufacturer's authorised representative in the EU for product safety is Oxford University Press España S.A. of el Parque Empresarial San Fernando de Henares, Avenida de Castilla, 2 – 28830 Madrid (www.oup.es/en or product.safety@oup.com). OUP España S.A. also acts as importer into Spain of products made by the manufacturer.

www.ingramcontent.com/pod-product-compliance
Lightning Source LLC
LaVergne TN
LVHW102008210926
848483LV00010B/172

* 9 7 8 0 1 9 8 5 6 5 3 0 7 *